Introduction to Immunology

11th Hour

Introduction to Immunology

Janet M. Decker
Department of Veterinary Science and Microbiology
University of Arizona
Tucson, Arizona

b
Blackwell
Science

Editorial Offices:
Commerce Place, 250 Main Street, Malden, Massachusetts 02148, USA
Osney Mead, Oxford OX2 0EL, England
25 John Street, London WC1N 2BL, England
23 Ainslie Place, Edinburgh EH3 6AJ, Scotland
54 University Street, Carlton, Victoria 3053, Australia
Other Editorial Offices:
Blackwell Wissenschafts-Verlag GmbH, Kurfürstendamm 57, 10707 Berlin, Germany
Blackwell Science KK, MG Kodenmacho Building, 7-10 Kodenmacho Nihombashi, Chuo-ku, Tokyo 104, Japan

Distributors:
USA
Blackwell Science, Inc.
Commerce Place
350 Main Street
Malden, Massachusetts 02148
(Telephone orders: 800-215-1000 or 781-388-8250; fax orders: 781-388-8270)

Canada
Login Brothers Book Company
324 Saulteaux Crescent
Winnipeg, Manitoba, R3J 3T2
(Telephone orders: 204-224-4068)

Australia
Blackwell Science Pty, Ltd.
54 University Street
Carlton, Victoria 3053
(Telephone orders: 03-9347-0300; fax orders: 03-9349-3016)

Outside North America and Australia
Blackwell Science, Ltd.
c/o Marston Book Services, Ltd.
P.O. Box 269
Abingdon
Oxon OX14 4YN
England
(Telephone orders: 44-01235-465500; fax orders: 44-01235-465555)

Acquisitions: Nancy Hill-Whilton
Development: Jill Connor
Production: Louis C. Bruno, Jr.
Manufacturing: Lisa Flanagan
Interior design by Colour Mark
Cover design by Madison Design
Typeset by Best-set Typesetter Ltd., Hong Kong
Printed and bound by Capital City Press

Printed in the United States of America
00 01 02 03 5 4 3 2 1

Library of Congress Cataloging-in-Publication Data

Decker, Janet M.
Introduction to immunology / by Janet M. Decker.
p. cm.—(11th hour)
ISBN 0-632-04415-2
1. Immunology—Outlines, syllabi, etc. 2. Immunology—Examinations, questions, etc.
I. Title. II. 11th hour (Malden, Mass.)
[DNLM: 1. Immunity—Programmed Instruction. 2. Allergy and Immunology—Programmed Instruction.
QW 518.2 D295i 2000]
QR181 .D395 2000
616.07'9—dc21 99-046471

CONTENTS

11th Hour Guide to Success vii
Preface viii

Unit I: Basic Concepts 1

1. Properties of Immunity 3
   1. Immune Recognition, Recall, and Protection 3
   2. Antigens and Immune Recognition 5
   3. Immune System Organization 8
   4. Immune Effector Mechanisms 10

2. Inducing, Detecting, and Investigating Immunity 16
   1. Inducing Immunity 16
   2. Detecting and Quantifying Molecules 18
   3. Detecting and Quantifying Cells 21
   4. Experimental Systems 23

Unit II: Antigen Recognition 31

3. Antibody Structure and Function 33
   1. Basic Antibody Structure 33
   2. Antibody Interaction with Antigen 35
   3. Functional Classes of Antibody 37
   4. Designer Antibodies 39

4. Antibody Gene Organization and Expression 45
   1. Organization of Immunoglobulin Genes 46
   2. Somatic Recombination 48
   3. B-Cell Development 50
   4. Isotype Switching 51

5. Major Histocompatibility Complex and Antigen Presentation 57
   1. MHC Proteins and Genes 57
   2. Antigen-Presenting Cells 60
   3. Antigen Processing and Presentation 62
   4. MHC and Immune Responsiveness 64

6. T-Cell Antigen Recognition 70
   1. TCR Structure and Antigen Recognition 70
   2. Producing a Functional TCR Molecule 72

3. Thymus Education of T Lymphocytes 74
4. Alloreactivity of T Lymphocytes 77

MIDTERM EXAM 82

Unit III: Immune Effector Mechanisms 87

7. Cytokines 89
1. General Properties of Cytokines 89
2. Cytokine Functions 91
3. Cytokine Receptors and Antagonists 93

8. Complement 100
1. Functions of Complement 100
2. Complement Cascades 103
3. Regulation of Complement Activity 105

9. Innate Immunity and Inflammation 111
1. Adhesion Molecules 113
2. Antigen Elimination 114
3. Innate and Early Induced Responses 115

10. Adaptive Immunity 121
1. T-Dependent and T-Independent Antigens 121
2. Activation of T Lymphocytes 123
3. Humoral Immunity 125
4. T-Cell–Mediated Immunity 128

Unit IV: Clinical Immunology 137

11. Hypersensitivity and Allergies 139
1. IgE-Dependent Hypersensitivity (Type I) 139
2. Antibody-Dependent Cytotoxicity (Type II) 141
3. Immune Complex–Dependent Hypersensitivity (Type III) 143
4. Delayed-Type Hypersensitivity (Type IV) 145

12. Autoimmunity and Immune Deficiencies 150
1. Immune System Regulation 150
2. Autoimmune Diseases 152
3. Immune Deficiencies 155

13. Manipulating the Immune System 162
1. Vaccines 162
2. Transplantation 165
3. Tumor Immunology 167

FINAL EXAM 173

Index 177

11TH HOUR GUIDE TO SUCCESS

The 11th Hour Series is designed to be used when the textbook doesn't make sense, the course content is tough, or when you just want a better grade in the course. It can be used from the beginning to the end of the course for best results or when cramming for exams. Both professors teaching the course and students who have taken it have reviewed this material to make sure it does what *you* need it to do. The material flows so that the process keeps your mind actively learning. The idea is to cut through the fluff, get to what you need to know, and then help you understand it.

Essential Background. We tell you what information you already need to know to comprehend the topic. You can then review or apply the appropriate concepts to conquer the new material.

Key Points. We highlight the key points of each topic, phrasing them as questions to engage active learning. A brief explanation of the topic follows the points.

Topic Tests. We immediately follow each topic with a brief test so that the topic is reinforced. This helps you prepare for the real thing.

Answers. Answers come right after the tests; but, we take it a step farther (that reinforcement thing again), we explain the answers.

Clinical Correlation or Application. It helps immeasurably to understand academic topics when they are presented in a clinical situation or an everyday, real-world example. We provide one in every chapter.

Demonstration Problem. Some science topics involve a lot of problem solving. Where it's helpful, we demonstrate a typical problem with step-by-step explanation.

Chapter Test. For more reinforcement, there is a test at the end of every chapter that covers all of the topics. The questions are essay, multiple choice, short answer, and true/false to give you plenty of practice and a chance to reinforce the material the way you find easiest. Answers are provided after the test.

Check Your Performance. After the chapter test we provide a performance check to help you spot your weak areas. You will then know if there is something you should look at once more.

Sample Midterms and Final Exams. Practice makes perfect so we give you plenty of opportunity to practice acing those tests.

The Web. Whenever you see this symbol the author has put something on the Web page that relates to that content. It could be a caution or a hint, an illustration or simply more explanation. You can access the appropriate page through *http://www.blackwellscience.com.* Then click on the title of this book.

The whole flow of this review guide is designed to keep you actively engaged in understanding the material. You'll get what you need fast, and you will reinforce it painlessly. Unfortunately, we can't take the exams for you!

PREFACE

This text was written to help you review your knowledge of basic immunology. It covers the material taught in most one-semester introductory immunology courses for undergraduate and health-care professional students. It lacks the detailed information and numerous examples of a complete textbook, many of which are available; but, it supplies what most immunology texts lack, a means of testing your knowledge of basic immunology concepts and your ability to apply this knowledge to real-life situations. The longer I teach, the more convinced I am that the amount of active participation by the student determines how much effective learning takes place. I hope your use of this text enhances your understanding and appreciation of a very exciting and important field. The availability of more complete information on the Internet is indicated by ; the web address is http://www.blackwellscience.com.

I would never have begun this book without the encouragement of Nancy Hill-Whilton, Executive Editor; I would never have finished without the gentle prodding of Jill Connor, Assistant Development Editor. My sincere thanks to both of them and to the talented production staff at Blackwell Science. I gratefully acknowledge the following professors and students for their manuscript reviews and helpful comments: Min-Ken Liao, Hope College; Shirley Kovacs, CSU Fresno; Patrick Sonner, Ohio Northern University; Emalie Carson, Ohio Northern University; Michael Misfeldt, University of Missouri; Robert Turner, Western Oregon University; and Chris Coverdill, Western Oregon University. I also thank my immunology students at the University of Arizona, who supplied the germ of many of these questions and who keep me excited about teaching.

This book is dedicated to Dom and Larissa.

Janet M. Decker, Ph.D.
Senior Lecturer
Veterinary Science and Microbiology
University of Arizona

UNIT I:
BASIC
CONCEPTS

Properties of Immunity

Long before the causes of infectious disease were understood, humans observed that surviving certain diseases conferred specific long-lasting protection. By the 19th century, humans systematically induced immunity by exposing themselves and their animals to weakened forms of infectious agents. Immunologists are still working to develop new vaccines and to control allergies, transplant rejection, and autoimmunity.

ESSENTIAL BACKGROUND

- **Basic eukaryotic and prokaryotic cell structure**
- **Role of enzymes in cell function**
- **Protein, carbohydrate, and nucleic acid structure**
- **Gene structure, function, regulation, and inheritance**

TOPIC 1: IMMUNE RECOGNITION, RECALL, AND PROTECTION

KEY POINTS

✓ *What are the properties of innate immunity?*

✓ *How is adaptive immunity different from innate immunity?*

✓ *How can immunity be acquired?*

The immune system is a collection of cells and organs that can recognize and destroy infectious organisms (**pathogens**). We are all born with certain natural (**innate**) protective mechanisms against infectious disease, including the physical and physiological barriers that prevent the entry of infectious agents such as skin, mucus, tears, saliva, respiratory tract cilia, stomach acid, and antibacterial agents such as lysozyme. Innate immunity also includes inflammation, pain, fever, coughing, vomiting, diarrhea, and healing. Innate immunity that protects us against pathogens in the tissues and circulation is mediated by **effector cells** called **phagocytes**, which engulf and destroy pathogens, and by **natural killer** (**NK**) cells, which recognize and destroy virus-infected cells without engulfing them. Phagocytic cells include two types of **leukocytes** (white blood cells): **monocytes**, also called **macrophages** when they leave the circulation and enter the tissues, and **polymorphonuclear leukocytes** (**PMNs** or **granulocytes**), including **neutrophils**, **eosinophils**, and **basophils**. Macrophages and PMNs release proteolytic enzymes, strong oxidants, and other bacteriocidal molecules. They also stimulate inflammation with vasoactive compounds that allow fluid movement from the circulation into the tissues and chemotactic molecules that attract leukocytes. Blood **complement** proteins promote pathogen phagocytosis and lysis and also innate immunity. Innate immunity begins when antigens enter the body and operates against many but not all pathogens. It is triggered by commonly shared molecules on pathogens and is not specific to a particular antigen. The speed and magnitude of the innate response is not affected by prior exposure to the antigen.

Acquired or **adaptive** immunity develops over several days or weeks after exposure to antigen and is antigen specific. Each exposure to an antigen induces a faster and greater adaptive immune response to that antigen, a process called **immune memory**. **Lymphocytes** are antigen-specific leukocytes responsible for adaptive immunity. Lymphocytes recognizing many diverse antigens are produced continually even in the absence of antigen exposure. When a lymphocyte encounters its specific antigen and receives the proper co-stimulatory signals, it proliferates and differentiates into a clone of effector cells with the same antigen specificity (**clonal selection**). **B lymphocytes** (B cells) respond by becoming **plasma cells** that secrete antigen-binding proteins called **antibodies**. Antibodies bind extracellular pathogens and bacterial toxins to inactivate them; they also **opsonize** (coat) the pathogens to promote phagocytosis. Some **T lymphocytes** (**T cells**) respond to intracellular pathogens like viruses by becoming **cytotoxic T lymphocytes** (**T_C**), which destroy infected cells. Other T lymphocytes become active **helper T cells** (**T_H**), which stimulate antibody synthesis and macrophage activation by secreting signaling molecules called **cytokines**. Both B and T lymphocytes also respond to antigen plus T_H cytokines by differentiating into **memory cells**, which are long-lived and respond more quickly than naive lymphocytes when reexposed to antigen.

Adaptive immunity is usually acquired actively by natural infection or by vaccination with killed or weakened (**attenuated**) pathogen or inactivated toxin. Adaptive immunity can also be acquired passively from an immune person by the transfer of antibodies or immune cells. Immunity that can be transferred in serum is called **humoral immunity**; examples are antibodies transferred across the placenta or in breast milk from mother to child and horse antibodies to rattlesnake venom used to treat snake bite. Immunity that can be transferred only with T-cell transfer is called **cellular immunity**; passive cellular immunity is limited by rejection of foreign transplanted cells and is usually only done between inbred animals or in human bone marrow transplants where the entire blood cell-forming system is transferred.

Topic Test 1: Immune Recognition, Recall, and Protection

True/False

1. Lymphocytes acquire their antigen specificity before they encounter antigen.
2. Both T and B lymphocytes perform cellular immunity.

Multiple Choice

3. An example of innate immunity is
   a. Antibody production by plasma cells.
   b. Antigen removal by cilia in the respiratory tract.
   c. Complement activation by antibody bound to the surface of a bacterium.

d. Memory response to influenza virus.
e. Recognition and killing of virus-infected cells by cytotoxic T cells.

4. Humoral immunity can be acquired passively by
   a. Catching a virus from a friend by shaking hands.
   b. Receiving a vaccine of influenza virus grown in eggs.
   c. Receiving serum from someone who has recovered from an infection.
   d. Receiving leukocytes from an immune family member.
   e. Sharing a soda with someone who has a cold.

Short Answer

5. How does one acquire innate immunity?
6. What is immune memory?

Topic Test 1: Answers

1. **True.** Antigen-specific B and T lymphocytes are produced continuously. Exposure of an immature lymphocyte to antigen may trigger its destruction; this mechanism prevents an immune response to self-antigens.
2. **False.** Both cells are responsible for adaptive immunity; however, cellular immunity is classified as responses mediated by T lymphocytes.
3. **b.** Innate immunity such as mechanical removal of antigen is not antigen specific or inducible. The other choices all involve antibody or cytotoxic T cells, which are antigen specific and represent adaptive immune responses.
4. **c.** Humoral immunity can be acquired passively by receiving serum from someone who has recovered from an infection. Answer d is passive cellular immunity; the other answers describe ways of acquiring active immunity.
5. Innate immunity is present from birth.
6. Immune memory is the ability to make a faster and greater immune response against an antigen after repeated exposure to that antigen. This is the biological basis for immunization.

TOPIC 2: ANTIGENS AND IMMUNE RECOGNITION

Key Points

✓ *What is the chemical nature of antigens?*

✓ *What properties increase a molecule's immunogenicity?*

✓ *How do lymphocytes recognize antigen?*

Antigens are molecules that generate the production of antibodies. In practice, the term antigen is used to mean any molecule recognized by the immune system. Antigens that induce adaptive immunity are called **immunogens**. All immunogens are antigens, but some antigens, called **haptens**, are not immunogens. Immunogenicity is determined by the chemical nature of the antigen, by its foreignness, and by the conditions under which the immune system encounters it. Immunogenicity is generally higher for proteins than for other organic molecules; it increases with molecular size and complexity. Haptens bind antibody and are generally nonproteins too small to be immunogenic unless they are covalently attached to a **carrier protein** molecule. Immunogenicity can also be increased by mixing the antigen with **adjuvants**, which are molecules that release the antigen slowly and stimulate phagocytosis.

Cells of the innate immune system are not antigen specific; they have molecules on their membranes that bind antigens found on many infectious agents or in adjuvants. Macrophages, PMNs, and NK cells also have membrane receptors for complexes of antibody with antigen (**Fc receptors**, **FcR**) or complement with antigen (**complement receptors**, **CR**), so that antigen that has bound antibody or complement is more easily engulfed.

Lymphocytes of the immune system bind antigen using antigen-specific membrane proteins. The antigen receptor on B cells (**BCR**) is antibody; each membrane BCR has two identical binding sites for antigen, and each B cell has about 10^5 BCR, all with identical antigen-binding sites. Antibodies bind (are specific for) protein and nonprotein antigens. Each part of the antigen recognized by an antibody is called an **epitope**; most proteins have several epitopes that are recognized by different B cells. Epitopes may be shared by closely related antigens (**cross-reactivity**). A protein epitope may be a linear sequence of amino acids or it may be a structural feature assembled by protein folding. Epitopes with definite three-dimensional shapes and charged amino acids are particularly well recognized by antibodies; external membrane and cell wall molecules, often present in many copies on the pathogen, are common B-cell antigens. A small peptide of four to six amino acids could fit into an antibody binding site, but larger proteins have more extended epitopes across their surfaces. BCR and soluble antibody bind antigen in its native conformation.

The T-cell antigen receptor (**TCR**) is structurally related to antibody. Each T cell has about 10^5 TCR that share the same antigen specificity, and each TCR binds a single antigen epitope. Most TCR bind only peptide epitopes 9 to 20 amino acids long; these peptides must be processed (cut from proteins) and presented on (bound to) **major histocompatibility complex** (**MHC**) molecules. MHC proteins are the tissue typing antigens that must be matched to prevent transplant rejection; their primary function is antigen presentation. **MHC class II** molecules are found predominantly on **antigen-presenting cells** (**APC**; e.g., dendritic cells, macrophages, and B cells) whose function is to present antigen to and activate T cells. T_H bind peptide on class II MHC of APC which have internalized extracellular antigens, such as bacteria. **MHC class I** molecules are present on membranes of all nucleated cells. When these cells become infected with viruses, they can present virus peptides on their MHC class I to the TCR of T_C cells; the cytotoxic T cells become activated to kill the infected cells (**targets**). Because TCR binds processed antigen, binding does not depend on natural antigen folding; it does depend on how efficiently the antigen can be processed, bind MHC, and bind TCR.

Some molecules on pathogens bind lymphocyte surface molecules that are not antigen receptors. If this binding induces the lymphocytes to undergo cell division (mitosis), the molecules are called **mitogens**. Mitogens usually induce proliferation in a high frequency of lymphocytes regardless of their antigen specificity, a process called **polyclonal activation**. Some mitogens are **T-independent antigens**, capable of inducing B cells to secrete antibody in the absence of T-cell help.

Topic Test 2: Antigens and Immune Recognition

True/False

1. Each lymphocyte has many antigen receptors, each receptor capable of binding a different copy of the same antigen epitope.
2. Macrophages have FcR, which give them the same antigen binding specificity as B cells.

Multiple Choice

3. The antibiotic penicillin is a small molecule that does not induce antibody formation. However, penicillin binds to serum proteins and forms a complex that in some people induces antibody formation, resulting in an allergic reaction. Penicillin is therefore
   a. An antigen.
   b. A hapten.
   c. An immunogen.
   d. Both a and b
   e. Both a and c
4. For specific antigen recognition by T cells,
   a. Antigen is bound by a T-cell membrane antibody.
   b. Denaturation of antigen does not reduce epitope recognition.
   c. Developing T cells must be exposed to the antigen.
   d. MHC molecules are not required.
   e. Soluble antigen is bound directly without processing.

Short Answer

5. Name the APCs and the cell types to which they present antigen.
6. What is a mitogen?

Topic Test 2: Answers

1. **True.** Collectively, lymphocytes recognize many different antigens, but all receptors on an individual lymphocyte have identical antigen-binding sites.
2. **False.** Macrophages do have FcR that bind antigen–antibody complexes, but the specificity of the receptors is for the antibody, not the antigen.
3. **d.** Penicillin is not immunogenic unless it is bound to a carrier molecule, which makes it a hapten. It is also an antigen, because it can bind antibody.
4. **b.** Antigens must be degraded to be presented, so denaturation of antigen before exposure would not prevent T-cell recognition.
5. B cells, macrophages, and dendritic cells present antigen to T_H lymphocytes.
6. A mitogen is a molecule that stimulates mitosis and induces polyclonal proliferation of lymphocytes. For example, bacterial lipopolysaccharide is mitogenic for all B cells.

TOPIC 3: IMMUNE SYSTEM ORGANIZATION

KEY POINTS

✓ *What is the source of the immune system cells?*

✓ *How are specific immune system cells identified?*

✓ *What are primary (central) and secondary (peripheral) lymphoid organs?*

Cells of the immune system develop in the bone marrow in a process called **hematopoiesis**; T lymphocytes complete their differentiation in the thymus. **Pluripotent** self-renewing stem cells divide and differentiate into all functional blood cell types. At each stage of differentiation, cells become more restricted in their potential than their progenitors. During hematopoiesis, lymphocytes acquire their specific antigen receptors, co-receptors required for response to antigen, cytokine receptors, and adhesion molecules that target the cells to particular immune organs. Hematopoiesis is regulated by growth factors, growth factor receptors, and programmed cell death (**apoptosis**).

Plasma membrane molecules (**markers**) are used to identify functional cell types that are indistinguishable microscopically. Membrane markers or **CD** (cluster of differentiation) antigens are detected by their ability to bind antibodies that recognize them. Mature B-cell markers include antigen-specific membrane antibody, MHC class II, FcR, CR2 (CD21), and CD19. Mature T-cell markers include antigen-specific TCR, CD2, CD3, and either CD4 on T_H (which binds MHC class II) or CD8 on T_C (which binds MHC class I). Cells of the innate immune system also have membrane markers. Many of these, like FcR and CR, are shared between cells; a few are lineage specific, like CD115, the receptor for a growth factor that specifically binds monocytes and macrophages. NK cells are large granular lymphocytes that share markers with both T cells and monocytes; they also have the NK-specific adhesion molecule CD56 and killer inhibitory receptors.

Cells of the immune system are present throughout the body. Clusters of lymphocytes and specialized antigen-collecting epithelial cells called **M cells** line the mucous membranes of the respiratory, digestive, and urogenital systems where the potential for contact with pathogens is highest. M cells, tonsils, appendix, and Peyer's patches are called the **mucosal associated lymphoid tissues**. Other **peripheral** lymphoid organs are the spleen and lymph nodes, where antigens from the blood and tissues encounter the immune system. Peripheral lymphoid organs are highly organized to facilitate interactions among antigen, APCs, and B and T lymphocytes. **Central** (**primary**) lymphoid organs for the antigen-independent production and maturation of lymphocytes and the removal of self-specific T and B cells include the bone marrow and the thymus. Fluid that bathes tissues transports nutrients and waste products into the lymph nodes, carrying antigen with it. Lymphatic vessels transport lymph and cells from the lymph nodes to the blood circulatory system. At any given time, many leukocytes circulate throughout the body and are present in high numbers in peripheral blood. Expression of adhesion molecules on endothelial cells lining blood vessels is increased by inflammatory cytokines and signals leukocytes to enter the tissues or the secondary lymphoid organs in response to antigen.

Topic Test 3: Immune System Organization

True/False

1. A mature T_H cell can be identified by the presence of CD4, CD8, and TCR.
2. Self-antigen–specific lymphocytes are produced in the primary lymphoid organs.

Multiple Choice

3. Peripheral lymphoid organs
   a. Are centrally located in the abdomen to protect their vital functions.
   b. Are designed to maximize contact between antigen and lymphocytes.
   c. Produce antigen-specific lymphocytes from stem cells in response to antigen.
   d. Sequester antigen to minimize its damage to the body.
   e. Store large numbers of activated effector cells for a rapid response to antigen.
4. Membrane markers on immune system cells
   a. Are present on lymphocytes but not other cells of the immune system.
   b. Are put there by immunologists so they can identify functional cell types.
   c. Attract antigen to the secondary lymphoid organs.
   d. Can be antigen receptors.
   e. Develop once the cells leave the primary lymphoid organs.

Short Answer

5. What cells and functions would be deficient in a child born without a thymus?
6. Why do we get swollen glands during an infection?

Topic Test 3: Answers

1. **False.** T_H and T_C cells are differentiated by the presence of CD4 on the former and CD8 on the latter; both have TCR and CD3.
2. **True.** The generation of lymphocytes can produce self-specific cells, which are killed before they leave the primary lymphoid organs.
3. **b.** Peripheral lymphoid organs are designed to maximize contact between antigen and the immune system, so they are distributed throughout the body.
4. **d.** Membrane markers can be antigen receptors, co-receptors, adhesion molecules, and other identifying molecules. Leukocytes acquire their markers during development but may change them upon activation.
5. A child born without a thymus would lack functional T cells: T_C to kill virus-infected cells and T_H to activate B-cell antibody secretion and macrophage killing of engulfed pathogens. Most adaptive immunity would suffer.
6. The swollen "glands" are lymph nodes, which enlarge because lymphocytes are retained during antigen contact and proliferate into clones of effector cells.

TOPIC 4: IMMUNE EFFECTOR MECHANISMS

KEY POINTS

✓ *What is the ultimate purpose of the immune system?*

✓ *How does the immune system respond to extracellular (exogenous) antigens?*

✓ *How does the immune system respond to intracellular (endogenous) antigens?*

When the immune system functions optimally, antigen is eliminated before the host develops symptoms. **Exogenous** (extracellular) antigens, including most bacteria and their toxins, are most accessible to immune elimination. Phagocytes bind commonly shared antigens on bacterial surfaces, engulf the bacteria, and digest them inside phagocytic vesicles. Digestive enzymes, oxygen radicals, and peroxides released by macrophages and PMNs during inflammation kill pathogens. Complement is activated by binding bacterial surface molecules; once activated, it binds phagocyte CRs to promote phagocytosis. Other complement products attract leukocytes and increase blood vessel leakiness so that complement, clotting factors, and antibody reach the infection site. In some cases, complement directly lyses the bacterium.

Once adaptive immunity is triggered, antibody is the primary tool for exogenous antigen elimination. Antibodies, also called **immunoglobulins** (Ig), are antigen-binding proteins that are divided into five classes (**isotypes**) based on their structures. The amino acid sequence of antibodies differs from one molecule to the next in the antigen-binding site (**variable region**) but does not differ significantly in the rest of the molecule (**constant region**). Each isotype has different biological functions once antigen is bound, and the isotypes have different numbers of antigen-binding sites per molecule. IgM and IgD are antigen receptors (BCR) on B cells. After antigen activation, B cells secrete IgM. Secreted IgM has 10 antigen-binding sites; it binds antigen and activates complement very efficiently but enters the tissues slowly because of its size. When B cells receive helper cytokine signals from T cells, they begin secreting IgG or IgA. IgG is the predominant antibody isotype in serum; it has two antigen-binding sites and can easily enter the tissues from the circulation. IgG activates complement and IgG–antigen complexes bind FcR; phagocytes then use both CRs and FcR to facilitate phagocytosis. IgA is made predominantly in mucosal lymphoid tissues; it is present in mucous secretions of the respiratory, digestive, and urogenital tracts and in breast milk. IgG and IgA binding neutralize virus and toxin activities by blocking host cell binding. IgE is made to helminth (worm) parasites and to environmental antigens by people who have allergies; IgE binds FcR on mast cells, which respond by releasing histamine.

Endogenous (intracellular) antigens, including viruses, protozoan parasites, and bacteria that survive phagocytosis, are not directly accessible to phagocytes, complement, or antibody and must be eliminated by destruction of infected host cells. Viruses that insert their own molecules into the host cell membrane before using it as an envelope are somewhat more vulnerable, because antibody can bind these viral proteins and trigger complement-mediated lysis of the host cell. Antibody on phagocyte and NK cell FcR binds membrane-expressed viral antigen and triggers host cell killing (**antibody-dependent cell-mediated cytotoxicity**). For intracellular parasites that do not express their antigens on the host cell membrane, immune recognition requires antigen peptide presentation on host cell MHC class I molecules and specific recognition by T_C or less-specific recognition by NK cells, followed by host cell lysis. Pathogens that survive the digestive process, which usually follows phagocytosis, and live in macrophage vesicles, where their peptides get presented on MHC class II, are eliminated by T_H activation of the macrophage to kill vesicular pathogens.

Topic Test 4: Immune Effector Mechanisms

True/False

1. Antigen-presenting cells include B cells, dendritic cells, and macrophages.
2. Complexes of antigen with antibody can stimulate inflammation.

Multiple Choice

3. Effector functions of complement include all of the following *except*
   a. Attracting phagocytes to the site of infection.
   b. Facilitating phagocytosis of complement-coated bacteria.
   c. Increasing blood vessel permeability to plasma proteins.
   d. Lysing bacterial cells.
   e. Presenting antigen to B cells.
4. Antigen epitopes recognized by ________ tend to be ________.
   a. B cells; highly accessible sites on the exposed surface of the antigen.
   b. Cytotoxic T cells; highly accessible sites on the exposed surface of the antigen.
   c. Complement; highly accessible sites on the surface of an APC.
   d. Helper T cells; associated with MHC class I on infected cells.
   e. Macrophages; associated with MHC class II on APCs.

Short Answer

5. Polio virus is a nonenveloped virus that infects intestinal epithelial cells. How can the adaptive immune system eliminate polio virus from the body?
6. Which effector mechanisms are classified as adaptive immunity?

Topic Test 4: Answers

1. **True.** APC present antigen on MHC class II to T_H.
2. **True.** Complement activation by antigen–antibody complexes attracts leukocytes to the site of infection and induces blood vessel leakiness.
3. **e.** BCR binds antigen directly. Complement activation produces chemoattractants, products that increase blood vessel leakiness, and a complex that can lyse some cells.
4. **a.** BCR binds native antigen, usually exposed epitopes on the surface of the antigen, as do complement and macrophages. T_C bind antigen peptide on infected cell MHC class I; T_H bind antigen peptide on APC MHC class II.
5. Because polio virus is a nonenveloped endogenous pathogen, its native proteins are not expressed on the epithelial cell membrane. Peptides are presented on MHC class I, and specific recognition and killing by T_C eliminates antigen.

6. Adaptive effector mechanisms are those involving antigen-specific recognition and memory responses: antibody secretion by plasma cells, cell lysis by T_C, and regulatory activities of T_H.

CLINICAL CORRELATION: TETANUS BOOSTERS

Clostridium tetani is a gram-positive rod-shaped bacterium that lives in soil. Under unfavorable growth conditions, it forms spores that can survive for long periods and then become vegetative cells when conditions are favorable. *C. tetani* grows only in the absence of oxygen, thriving in deep wounds and living outside host cells. It secretes a neurotoxin that causes the muscle spasms that give tetanus its other name, "lockjaw." Infants are injected with tetanus toxoid, an inactivated *C. tetani* exotoxin, several times during their first 2 years after birth. A toxoid is used because the toxin is deadly at doses too low to induce immunity; antibodies made against the toxoid neutralize the cross-reactive toxin. Booster vaccinations are recommended every 10 years to maintain immunity. In cases where active immunity is uncertain, passive immunization with serum from vaccinated humans may be given to quickly neutralize toxin. Passive immunization has the same half-life as human IgG, about 21 days, so active immunization is given at the same time to provide longer-lasting protection.

DEMONSTRATION PROBLEM

Influenza virus infects respiratory epithelial cells. It is an enveloped virus, and virus hemagglutinin (H) and neuraminidase (N) proteins, which help the virus enter and leave the host cells, are expressed on host cell plasma membranes that envelope the emerging virus particles. Influenza viruses rapidly mutate their H and N antigens, so that exposure or immunization to influenza one year does not guarantee immunity the next. "Flu shots" contain virus particles grown in fertilized eggs and inactivated so they do not infect vaccine recipients. The vaccine also contains adjuvants that promote virus phagocytosis. Describe the immune effector mechanisms generated by flu shots and how they prevent influenza when an immunized person is exposed.

SOLUTION

Because the influenza virus in flu shots cannot infect host cells, it behaves as an exogenous antigen and induces humoral immunity. Virus binds to BCR specific for H and N protein epitopes. Adjuvant stimulates macrophages to phagocytose and present H and N peptides on MHC class II to T_H, which secrete cytokines to stimulate B-cell activation, antibody secretion, and memory cell production. Antigen-specific T_C are not activated by the vaccine. Upon infection with live influenza virus, antibody blocks virus binding to host cells, opsonizes influenza particles to facilitate phagocytosis, and activates complement to attract more phagocytes to the respiratory tract.

Chapter Test

True/False

1. Recognition and killing of virus-infected cells by T_C cells is adaptive cellular immunity.
2. An example of active humoral immunity is treatment with horse anti-snake venom.
3. Each lymphocyte has many antigen-binding receptors, each receptor capable of binding a different antigen epitope.
4. Peptidoglycan (a bacterial polymer of *N*-acetylglucosamine and *N*-acetylmuramic acid on a protein backbone) is more immunogenic than glycogen (a polymer of glucose).
5. NK cells are a part of innate cellular immunity.

Multiple Choice

6. Which statement about antigen epitopes is *false*?
   a. An epitope may be shared by two different antigens.
   b. A protein molecule usually contains multiple epitopes.
   c. B cells bind only processed antigen epitopes.
   d. Epitopes may be linear or assembled.
   e. Some epitopes are more immunogenic than others.
7. Hematopoietic stem cells are pluripotent, which means that they
   a. Are antigen-specific cells.
   b. Are capable of developing into any blood cells.
   c. Are committed to produce cells of a single lineage.
   d. Are not self-renewing.
   e. Will develop into T and B lymphocytes of many different antigen specificities.
8. Phagocytosis
   a. Can be stimulated by antigen binding to complement or antibody.
   b. Is an antigen-specific process.
   c. Must be preceded by antigen processing.
   d. Rids the body of virus-infected cells.
   e. Only occurs after plasma cells begin secreting antibody.
9. Primary lymphoid organs
   a. Are efficient in exposing T cells to foreign antigen presented on dendritic cells.
   b. Are the primary site of antibody synthesis and release.
   c. Filter blood and trap bloodborne antigens.
   d. Provide the microenvironment for maturation of T and B cells.
   e. Line the mucosal surfaces of the body for efficient antigen contact.
10. Jenner observed that milkmaids who were infected with cowpox were later immune to smallpox infections. This is an example of a(n)
    a. Acquired immunity of barrier skin cells.
    b. Active immunization with a nonrelated organism that causes similar symptoms.
    c. Innate immunity of milkmaids to smallpox.
    d. Memory response to a cross-reactive antigen.
    e. Passive immunization from contact with cow's milk antibodies.

Short Answer

11. How are antigen receptors on lymphocytes different from those on phagocytes?
12. Why can't inflammation successfully eliminate all infectious organisms?
13. Can cellular immunity be passively transferred? What are the limitations?
14. Why don't we vaccinate against all microorganisms?

Essay

15. How does clonal selection explain the adaptability, antigen specificity, discrimination between self and non-self, and memory characteristics of adaptive immunity?

Chapter Test: Answers

1. **T** 2. **F** 3. **F** 4. **T** 5. **T** 6. **c** 7. **b** 8. **a** 9. **d** 10. **d**

11. Antigen receptors on lymphocytes are of a single type and are antigen specific. Antigen is bound to phagocytes by a variety of molecules, including CRs and FcR, which are not antigen specific.
12. Inflammation cannot successfully eliminate all infectious organisms because organisms have evolved ways of avoiding the innate immune system.
13. Cellular immunity can be passively transferred with a transfusion of memory T cells; this works well only in genetically identical individuals where the T cells are not rejected.
14. Many microorganisms are not pathogenic; many others can be eliminated successfully by our immune systems or do not cause fatal disease. In other cases it has not been practical to vaccinate against numerous antigenic variants or a protective response has not been induced.
15. Primary lymphoid organs constitutively produce large numbers of lymphocytes with diverse antigen specificities; self-reactive T and B cells are killed in the thymus and marrow. Each lymphocyte is specific for a single-antigen epitope, and only lymphocytes specific for the stimulating antigen are triggered to divide into clones of effector cells. Specific memory cells are also produced in response to antigen; these more numerous, and longer lived cells produce a faster and larger response to subsequent contact with the same antigen, the basis of immune memory.

Check Your Performance:

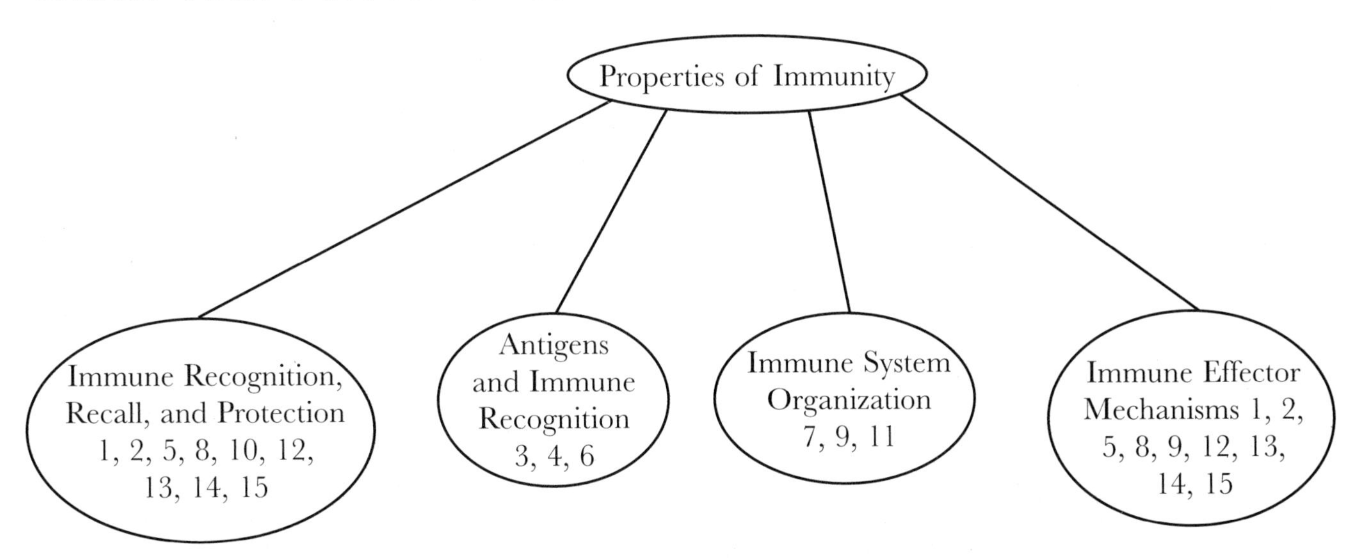

Check your understanding of this chapter by noting the number of questions for each topic you missed on the chapter test.

Inducing, Detecting, and Investigating Immunity

Immunologists have developed techniques for measuring innate and adaptive immunity and model systems in which they can experimentally vary elements of the immune system.

ESSENTIAL BACKGROUND

- **Gene expression and enzyme function**
- **Antigen structure and immunogenicity (Chapter 1)**
- **Immune effector cells and molecules (Chapter 1)**

TOPIC 1: INDUCING IMMUNITY

KEY POINTS

✓ *What factors are critical for inducing an immune response?*

✓ *How is a secondary response different from a primary response?*

✓ *What factors determine whether humoral or cellular immunity is induced?*

Active immunity is induced by antigen exposure. Antigen dose, chemistry, and immunogenicity and the route and timing of antigen contact influence the magnitude and nature of the immune response. Very high or low antigen doses induce **tolerance**, an unresponsiveness to that antigen, whereas intermediate doses induce immunity. Aggregated proteins are easily phagocytosed and more immunogenic than soluble proteins. Protein antigens that activate helper T cells (T_H) induce IgG or IgA synthesis and memory B- and T-cell generation; some carbohydrate antigens stimulate B-cell IgM secretion in the absence of T-cell help but do not elicit IgG or memory cells. Human infants make poor responses to carbohydrate antigens, so early vaccinations with bacterial polysaccharide antigens use protein carriers. Adjuvants, including bacterial products, alum, or oil emulsification, attract and activate macrophages to improve immunogenicity.

Antigens encountered in body tissues through subcutaneous or intramuscular injection are carried by lymph and macrophages into the lymph nodes, where they usually elicit production of serum IgG antibody. Antigens encountered by mucosal routes (in the respiratory, digestive, or urogenital tracts) are engulfed by M cells and induce secretory IgA production. Oral tolerance to foods is common, although some foods induce allergic reactions. IgE is often made to worm parasites and to protein allergens encountered at low doses. Introducing antigens into the circulation requires larger doses for inducing immunity, because most antigen is quickly removed and destroyed in the spleen and liver reticuloendothelial systems. Exogenous (extracellular) pathogens and their toxins induce formation of antibodies. Endogenous (intracellular) pathogens elicit formation of cytotoxic T lymphocytes (T_C) to kill virus-infected cells or inflammatory cytokine-producing T_H lymphocytes to promote macrophage destruction of engulfed pathogens. A virus must be capable of infecting and replicating in a host cell to induce formation of T_C, but both attenuated (weakened) and killed viruses induce antibody formation.

Upon first (primary) contact with antigen, there is a lag of several days before increased antibody production or cellular immunity is detected. During this lag, the innate immune systems are responding: macrophages and neutrophils engulf and destroy the pathogens, complement is activated to stimulate inflammation and pathogen lysis, natural killer (NK) cells kill virus-infected host cells, and cytokine production results in fever and increased hematopoiesis. Antigen is carried to secondary lymphoid organs where it stimulates lymphocytes to become immune effector cells. If antigen is not removed by innate immune mechanisms, adaptive immune effector functions increase until about 10 to 14 days after antigen contact, plateau for several days, and, when antigen has been successfully eliminated, drop but remain higher than in unimmunized individuals. The predominant antibody made during the primary immune response is IgM. Upon secondary antigen contact, the lag period is shorter, the peak response higher, and increased antibody levels may persist for weeks or months. IgG (or IgA for mucosal antigens) is the predominant antibody made during secondary immune responses.

Topic Test 1: Inducing Immunity

True/False

1. Adjuvants increase immunogenicity by attracting and activating macrophages.
2. If an antigen is eliminated quickly by innate immune effector mechanisms, no adaptive immunity will develop.

Multiple Choice

3. Lymphocytes are activated by antigen in the
   a. Bloodstream.
   b. Bone marrow.
   c. Liver.
   d. Lymph nodes.
   e. Skin.
4. During the lag period between antigen contact and detection of adaptive immunity,
   a. Antigen is hidden from the immune system in macrophages.
   b. Cellular immunity can be detected but antibodies cannot.
   c. Innate immune effectors are eliminating antigen.
   d. Innate immunity blocks the activation of adaptive immune effector cells.
   e. New B and T cells with the appropriate antigen specificity must be produced in the bone marrow.

Short Answer

5. How are the immune responses to endogenous and exogenous pathogens different?
6. How are the humoral immune responses to a primary and secondary contact with the same antigen different?

Topic Test 1: Answers

1. **True.** Adjuvants attract and activate antigen-presenting cells (APC), which then induce T-cell help.
2. **True.** Rapid elimination of antigen by phagocytosis, complement activation, or NK cells removes the antigen stimulus for lymphocytes.
3. **d.** Lymphocytes are activated by antigen in the secondary lymphoid tissues, including lymph nodes, spleen, and mucosal lymphoid areas.
4. **c.** During the lag period, macrophages, neutrophils, NK cells, and complement are eliminating antigen. Adaptive immune responses require time for lymphocytes to bind antigen, divide, and differentiate into effector cells.
5. Exogenous pathogens (most bacteria) activate T_H and B cells to elicit antibody production; antibody binds the bacteria to promote phagocytosis and complement-mediated lysis. Endogenous pathogens (viruses and some bacteria) activate T_C cell killing of infected host cells and T_H-stimulated macrophage killing of pathogens living in phagosomes. Antibody is also made to endogenous antigen.
6. A primary response to a protein antigen has a lag period of about 7 days, a peak IgM response in about 10 to 14 days, and a significant but small increase in serum antibody levels after antigen is completely removed. A secondary response to the same antigen has a lag of only 3 to 4 days, a peak IgG (or IgA) response in about a week, and a longer and higher plateau of increased serum antibody after antigen is eliminated.

TOPIC 2: DETECTING AND QUANTIFYING MOLECULES

KEY POINTS

✓ *What properties of antibodies are important for their use in each technique?*

✓ *What detection systems are used to enhance antibody measurements?*

✓ *Which techniques are best for detecting, quantifying, or purifying molecules?*

The ability of antibody to bind specific antigen allows antibody detection and quantification. It also makes antibody a useful tool for detecting, quantifying, and isolating other molecules to which specific antibody can be made.

IgM antibodies have 10 identical antigen-binding sites, although some sites may be blocked by the presence of antigen on nearby sites. When IgM binds to bacteria or red blood cells (RBC), it cross-links two adjacent cells, causing **agglutination** (clumping), which can be detected visually. IgG is too small to cross-link cells, which because of their negative charge are difficult to bring into close contact, unless a second anti-IgG antibody is added. When serum that may contain specific antibody is added to pure antigen, agglutination demonstrates the presence of antibody. The **titer** (amount) of antibody is measured by doing replicates using a constant amount of antigen and increasing twofold dilutions of serum. The highest dilution that gives a positive response is the antibody titer; for example, 1/4 is a very low titer (few specific antibody molecules are present), whereas 1/1,024 is a very high titer. To use agglutination to detect antigen, pure antibody is added to a sample that may contain antigen; if agglutination occurs, the specific antigen is present in detectable quantities. Agglutination is limited by the size and amount of antigen and is relatively insensitive but useful for identifying the **serotype** (antigenic specificity) of bacteria and blood type.

Smaller antigens may also be cross-linked by antibody. If the antigen/antibody ratio favors each antibody binding two antigens and each antigen binding at least two antibodies (equivalence), a visible precipitate is formed; less precipitate results at antibody or antigen excess. In **Ouchterlony double-diffusion assay**, antigen and antibody are put into small wells in agar; they diffuse through the agar and where they meet at equivalence they form a precipitate. Ouchterlony detects antigen or antibody and also assesses antigen similarity. When antibody is incorporated into the agar and antigen is added to wells, the diameter of the precipitate indicates antigen level. Precipitation also detects proteins separated by charge in **immunoelectrophoresis**. Precipitation is relatively insensitive and requires multivalent antigen and antibody.

ELISA is a much more sensitive technique for detecting antigen or antibody. To detect antibody, purified antigen is bound to a plastic surface; serum containing antibody is allowed to bind, followed by anti-IgG covalently bound to enzyme. After washing, the enzyme substrate that yields a colored product is added; color development indicates presence of specific antibody in the serum. The assay also quantifies antibody or antigen. Because of its sensitivity, safety, and ability to be automated, ELISA is the most commonly used immunoassay. **Radioimmunoassay** (**RIA**) is very similar to ELISA but uses radioactively labeled antigen or antibody.

Immunoblotting uses the ability of most antigens and antibodies to stick tightly to nitrocellulose membranes while retaining specific binding activity. In dot blots, small drops of antigen are put on nitrocellulose membrane and allowed to dry. Nonspecific binding is blocked and the membrane is incubated with antigen-specific enzyme-labeled antibody or with antigen-specific antibody and then enzyme-linked anti-IgG. After washing, the membrane is incubated in the enzyme substrate to allow colored product development. Proteins that have been separated by size using polyacrylamide gel electrophoresis (PAGE) or charge using isoelectric focusing are transferred to nitrocellulose membranes and reacted with specific antibody, a technique called **Western blotting**. Immunoblotting is sensitive and when combined with electrophoresis can be used to determine physical properties of the antigen.

Immobilized antibodies (usually bound to plastic beads) are used in **affinity chromatography** to purify molecules, including other antibodies. A mixture of proteins is passed over an affinity column and washed to remove unbound molecules. Specifically bound antigen is then eluted using low pH or chaotropic agents, molecules that disrupt the noncovalent binding between antigen and antibody.

Complement fixation detects antigen–antibody complexes and measures complement activity and relies on the fact that only antigen-bound antibody activates complement. Serum being tested for the presence of antibody is heated to 56°C for 30 minutes to inactivate complement (so that the amount of complement added to the test is controlled) before being added to antigen; heating does not inactivate antibody. Fresh complement is added that binds antigen–antibody complexes and then an indicator system of RBC plus anti-RBC is added. If all the complement has bound the antigen–antibody complex, none will be available to lyse the RBC (**hemolysis**); absence of lysis is a positive test. Complement-mediated hemolysis is used to detect antibodies to RBC, for example, those made by an Rh-negative woman against the Rh-positive RBC of her fetus (Coombs test).

Topic Test 2: Detecting and Quantifying Molecules

Multiple Choice

True/False

1. To form a precipitate, both antigen and antibody must have at least two binding sites.
2. Sue's antibody titer of 1/128 to polio virus indicates that she has more antibody than Sam's titer of 1/512.
3. ELISA
   a. Detects agglutination of cells by antibodies specific for cell-surface antigens.
   b. Uses isotope-labeled antibodies to detect extremely small amounts of antigen.
   c. Is used to isolate cell populations or proteins (including antibodies).
   d. Uses enzyme-linked anti-IgG to detect serum antibody to specific antigens.
   e. None of the above
4. Affinity chromatography *cannot* be used to
   a. Purify antibody by binding it to antigen.
   b. Purify antigen by binding it to antibody.
   c. Separate antibodies by their size.
   d. Separate antibodies to *Staphylococcus* from other serum antibodies.
   e. Separate IgM from IgG.

Short Answer

5. What properties of antibody make serological assays so useful?
6. ELISA is done as a screening test for human immunodeficiency virus (HIV); the antigen used is whole HIV grown in tissue culture. A positive ELISA is confirmed with a Western blot using solubilized HIV proteins. What information does Western blot provide that ELISA does not?

Topic Test 2: Answers

1. **True.** To form a precipitate, both antigen and antibody must have at least two binding sites so that large insoluble complexes can be formed.
2. **False.** Sam has more anti-polio virus antibody because his serum can be diluted fourfold more and still give a positive reaction.
3. **d.** Answer a is agglutination, b is RIA, and c is affinity chromatography.
4. **c.** Separating antibodies by their size is done using PAGE.

5. Antibodies bind antigen specifically and can be made to nearly any molecule. Antibodies can also be labeled or coupled to solid surfaces without impairing antigen binding. The ability of some antibody isotypes to activate complement, precipitate, or agglutinate antigens increases the range of detection systems.
6. ELISA detects serum antibodies to HIV in the infected person and also antibodies that react with non-HIV tissue culture proteins. Western blot detects antibodies to specific HIV proteins, separated by electrophoretic mobility.

TOPIC 3: DETECTING AND QUANTIFYING CELLS

KEY POINTS

✓ *How can the presence of cell surface markers be detected and quantified?*

✓ *How can cell populations be purified?*

✓ *How can functions of immune system cells be measured?*

To study immune function, it is often desirable to have pure populations of functional cells. Mixtures of lymphocytes, granulocytes, and macrophages can be separated from one another by physical properties like adherence or density. Macrophages adhere better to plastic at 37°C than lymphocytes; B cells bind more tightly to nylon wool than T cells. Mononuclear cells (mostly lymphocytes) are separated from whole blood by spinning blood layered onto Ficoll Hypaque in an analytical centrifuge; mononuclear cells remain at the interface and the more dense RBC and granulocytes pellet. These separations enrich or deplete but do not achieve purity.

Cells of the immune system have distinctive membrane markers that can be detected by antibody binding. Antibody is labeled with a fluorochrome (emits light when excited by ultraviolet radiation), radioisotope, enzyme, or gold. Bound antibody can be detected using a fluorescence microscope or a flow cytometer (**immunofluorescence**), photographic film that exposes in the presence of the radioisotope (**autoradiography**), a light microscope to detect the colored product of enzyme activity (**immunohistochemistry**), or an electron microscope to detect electron-dense gold (**immuno-electron microscopy**). Antibody cannot penetrate live cells, so only surface molecules will be detected unless the cells are permeabilized.

Cells with certain markers may be identified or eliminated (negatively selected) by incubating them with specific antibody and complement; complement-damaged cells are detected by their inability to exclude viability stains like trypan blue. Affinity chromatography is also used to separate cell populations; cells are incubated with immobilized antibody, nonbinding cells are washed off, and bound cells are eluted. Another simple and fast technique for separating cells uses marker-specific antibodies attached to paramagnetic beads; a strong magnet against the side of the tube retains cells bound to the beads, whereas nonbinding cells are washed away.

Fluorochrome-tagged antibody and **flow cytometry** detect and quantify cell populations. Suspensions of fluorescent antibody-tagged cells are passed in a narrow stream of droplets past a source of ultraviolet radiation (laser) and a detector. Side scatter of the light measures cell size and granularity, whereas emitted fluorescence measures the amount of marker present on the cell surface. A **fluorescence-activated cell sorter** (**FACS**) adds an electric charge to droplets containing the desired cell populations and deflects those cells into separate containers. Flow cytometry is slower than affinity chromatography or magnetic separation. Mixtures of cells are usually separated to more than 90% purity with affinity techniques and then to at least 99% purity by flow cytometry.

Cell function is also a measure of immunity. Antibody secreted by B cells and cytokines from T cells can be detected in serum or culture supernatants by ELISA. Antigen binding to B cells can be detected directly using labeled antigen. Because T cells only bind peptides presented on major histocompatibility complex (MHC), direct binding (which is very weak) is difficult to detect; antigen binding by T cells is usually measured by proliferation or cytokine production. Cell division requires DNA synthesis, which is quantified by ^{3}H-thymidine incorporation. Cytokines are detected directly by ELISA or indirectly by their effects on other cells, for example, inducing ^{3}H-thymidine incorporation by a cytokine-dependent cell line. T_H activation is also measured by stimulation of macrophage cytokine production or B-cell antibody secretion.

Mixed lymphocyte reactions are used to measure T-cell responses to foreign MHC antigens. Stimulator cells from the donor (treated to prevent or arrest cell division) are mixed with responder T cells from the recipient. Several days later the culture is pulsed with ^{3}H-thymidine and its uptake by responder cells is measured. The higher the radioactivity incorporated into the T cells, the stronger the response to the foreign cells and the more likely rejection will occur. **Cell-mediated lysis** is measured by **^{51}Cr release** from target cells. Live cells take up and retain ^{51}Cr; when they are incubated with effector T_C or NK cells, killing (percent specific lysis) is measured by ^{51}Cr release into the culture supernatant.

With increased ability to detect and measure specific nucleic acid sequences, many cell activities are measured by the presence of specific mRNA. Commonly used molecular techniques include polymerase chain reaction and reverse transcriptase polymerase chain reaction to increase the amount of nucleic acid and hybridization to identify the nucleic acid sequence.

Topic Test 3: Detecting and Quantifying Cells

True/False

1. The viability stain trypan blue is taken up by live cells but excluded by dead cells.
2. Complement activation by anti-CD3 can be used to positively select T cells from a mixture of T and B cells.

Multiple Choice

3. T cells and B cells are *not* different in their
   a. Ability to adhere to nylon wool columns.
   b. Effector functions.
   c. Membrane markers.
   d. Microscopic appearance.
   e. mRNA synthesis.
4. Activation of T_H *cannot* be measured by
   a. ELISA detection of antibody synthesis by B cells in response to T_H cytokines.
   b. ELISA detection of cytokines secreted by the activated T cells.
   c. Incorporation of ^{3}H-thymidine into DNA of the activated T cells.

d. Incorporation of ^{3}H-thymidine into DNA of an interleukin (IL)-2–dependent cell line incubated with culture supernatant from the activated T cells.
e. Release of ^{51}Cr from targets killed by the activated T cells.

Short Answer

5. Why can't antigen binding by T cells be detected by incubating the cells with labeled antigen?

6. How are MHC antigens detected for tissue typing?

Topic Test 3: Answers

1. **False.** The viability stain trypan blue is excluded by live cells but enters dead cells.

2. **False.** Complement activation by antibody to cell surface CD3 negatively selects T cells; because it kills the T cells, it cannot positively select them.

3. **d.** T cells and B cells are *not* different in their microscopic appearance.

4. **e.** Cytotoxicity is a function of T_C; the other assays either directly or indirectly measure T_H function.

5. T-cell receptor binds only peptides presented on MHC; binding of peptides alone or unprocessed protein is too weak to be easily detected.

6. Specific MHC antigens can be detected by complement-mediated killing in the presence of specific antibodies to MHC or via flow cytometry. Differences in MHC antigens between donor and recipient cells can be detected by mixed lymphocyte reaction or ^{51}Cr release.

TOPIC 4: EXPERIMENTAL SYSTEMS

KEY POINTS

✓ *How have some naturally occurring lymphoid tumors and immune deficiencies helped us understand the immune system?*

✓ *What are monoclonal antibodies and why are they useful?*

✓ *How do inbred, transgenic, and knock-out mice help us study the immune system?*

Much of our initial understanding of immune function came from investigating infection, lymphoid tumors, and inborn immune deficiencies. **Myelomas** (plasma cell tumors) arise from single plasma cells, produce homogenous antibody, and have allowed immunochemists to characterize Ig structure and function. Children born with specific deficiencies in their immune systems allowed physicians to identify functions of immune organs and cells. In strains of **nude** mice, impaired epithelial development results in absence of both hair and a functional thymus. Nude mice have normal macrophages, granulocytes, NK cells, B cells, and innate immunity and can produce IgM to some antigens and perform natural killing of virus-infected cells. They are very susceptible to infections and cannot reject tumors or skin grafts.

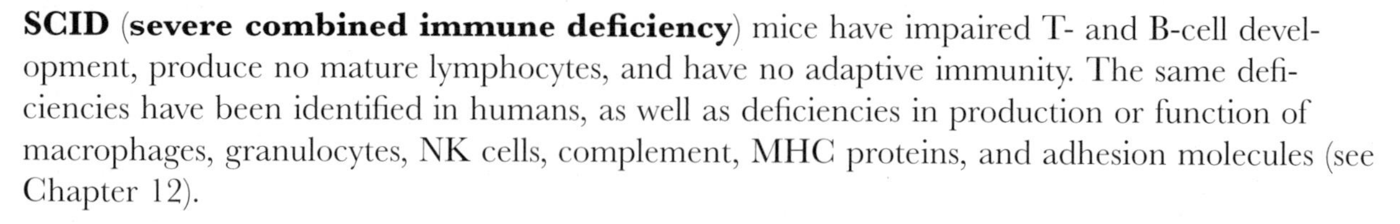

SCID (**severe combined immune deficiency**) mice have impaired T- and B-cell development, produce no mature lymphocytes, and have no adaptive immunity. The same deficiencies have been identified in humans, as well as deficiencies in production or function of macrophages, granulocytes, NK cells, complement, MHC proteins, and adhesion molecules (see Chapter 12).

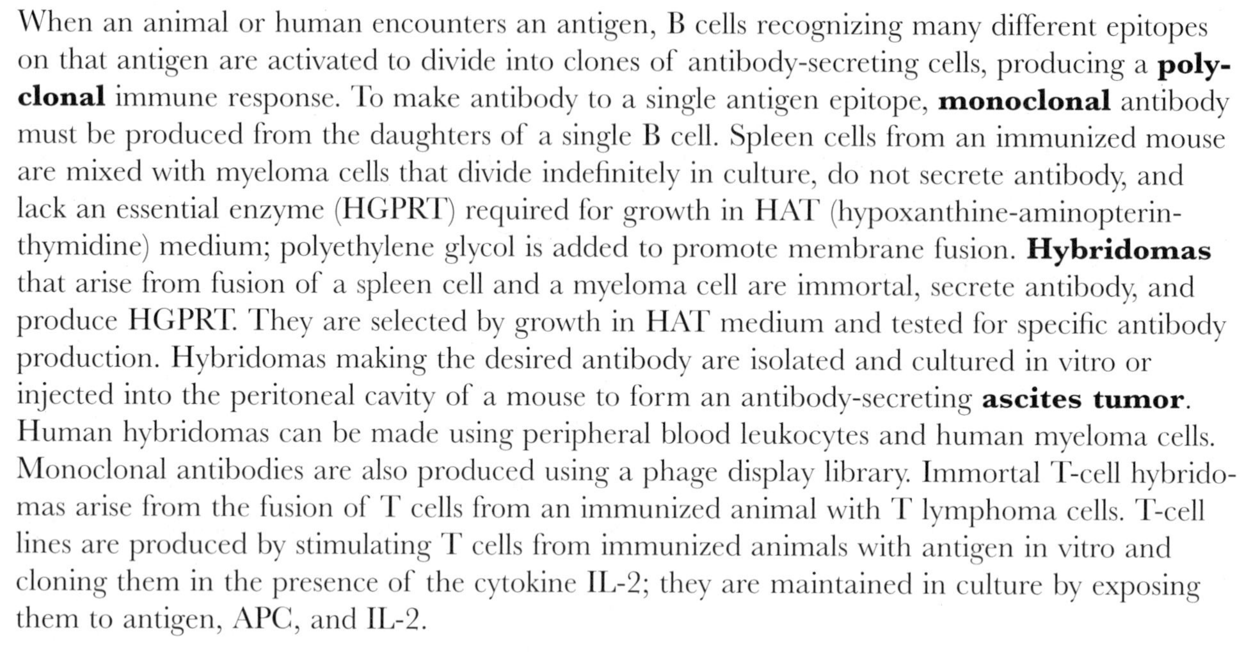

When an animal or human encounters an antigen, B cells recognizing many different epitopes on that antigen are activated to divide into clones of antibody-secreting cells, producing a **polyclonal** immune response. To make antibody to a single antigen epitope, **monoclonal** antibody must be produced from the daughters of a single B cell. Spleen cells from an immunized mouse are mixed with myeloma cells that divide indefinitely in culture, do not secrete antibody, and lack an essential enzyme (HGPRT) required for growth in HAT (hypoxanthine-aminopterin-thymidine) medium; polyethylene glycol is added to promote membrane fusion. **Hybridomas** that arise from fusion of a spleen cell and a myeloma cell are immortal, secrete antibody, and produce HGPRT. They are selected by growth in HAT medium and tested for specific antibody production. Hybridomas making the desired antibody are isolated and cultured in vitro or injected into the peritoneal cavity of a mouse to form an antibody-secreting **ascites tumor**. Human hybridomas can be made using peripheral blood leukocytes and human myeloma cells. Monoclonal antibodies are also produced using a phage display library. Immortal T-cell hybridomas arise from the fusion of T cells from an immunized animal with T lymphoma cells. T-cell lines are produced by stimulating T cells from immunized animals with antigen in vitro and cloning them in the presence of the cytokine IL-2; they are maintained in culture by exposing them to antigen, APC, and IL-2.

Limiting dilution studies are used to determine the frequency of antigen-specific cells in various cell populations. For example, dilutions of T cells are exposed to antigen plus APC, and the percent of cultures responding is measured. A statistical formula called the Poisson distribution can be used to calculate the frequency of responding cells if one responding cell is present per well when 37% of the wells are negative.

Selective breeding has produced **inbred mouse strains** used extensively to study immunity. Brother–sister matings are done until skin grafts between mice are accepted; this selects for mice that are homozygous for their MHC genes, which encode antigen-presenting molecules and therefore influence the ability of mice to make immune responses. **Congenic** mice that differ at a single MHC locus are bred to study the influence of individual MHC genes on graft rejection, antigen presentation, and antigen responsiveness. Inbred mouse strains are used for **adoptive transfer experiments**; mice are irradiated to destroy their own immune systems and then reconstituted using cells from immunized mice or cell lines of known antigen specificity. These **radiation chimeras** allow immunologists to do in vivo experiments with various cell populations. The ability to manipulate DNA has led to the production of **transgenic mice** in which a specific gene has been inserted and to **knock-out mice** in which a specific gene has been deleted.

Topic Test 4: Experimental Systems

True/False

1. Radiation chimeras are analogous to human bone marrow transplant recipients.
2. SCID mice can make innate immune responses to antigen.

Multiple Choice

3. Nude mice are deficient in their ability to produce
   a. Mature B cells.
   b. Mature T cells.
   c. NK cells.
   d. Pluripotent stem cells.
   e. None of the above

4. IL-4 is a cytokine produced by activated T cells of all mouse strains. To study the function of IL-4, you would
   a. Do brother–sister matings to develop a mouse strain homozygous for the IL-4 gene.
   b. Make a knock-out mouse for IL-4 and then see what immune deficiencies it has.
   c. Make a monoclonal antibody to IL-4 and use it to activate T cells.
   d. Make a mouse transgenic for the IL-4 gene from another mouse strain.
   e. Wait for an IL-4 mutation to occur in your mouse colony.

Short Answer

5. What are the steps that go into producing a monoclonal antibody to antigen X?

6. Some human infants are born with SCID due to a genetic deficiency in an enzyme (ADA) required for T- and B-cell development. How might you restore immune function to SCID infants?

Topic Test 4: Answers

1. **True.** Human bone marrow transplant recipients are irradiated to eliminate their own hematopoietic system, which has been replaced by stem cells from the bone marrow donor that can differentiate to form all the blood cells.

2. **True.** SCID mice cannot make adaptive immune responses to antigen.

3. **b.** Nude mice cannot produce mature T cells because their thymus is abnormal.

4. **b.** Because IL-4 is produced by all mouse strains, mice are already homozygous for the IL-4 gene. Adding more copies of the gene in a transgenic mouse will probably not tell you as much about its function as deleting the gene in a knock-out mouse and looking at what the mouse cannot do. Monoclonal anti–IL-4 could be used to block IL-4 function, but adding enough to bind all IL-4 in vivo is difficult. Waiting for a specific mutation to occur will take too long—and how will you recognize it when it happens?

5. Spleen cells from a mouse immunized with antigen X are mixed with myeloma cells and polyethylene glycol. Hybridomas are selected for their ability to grow in HAT medium and secrete anti-X. They are cloned so that all daughter cells come from one hybridoma and all antibody molecules secreted by the hybridoma are identical (monoclonal). The hybridoma is then cultured in vitro or in vivo to allow for collection of antibody.

6. A bone marrow transplant with stem cells possessing the missing enzyme or inserting a functional gene for the enzyme in a pluripotent stem cell from the SCID infant would reconstitute her or his immune systems.

DEMONSTRATION PROBLEM

Which assay(s) would you use for the following purposes? Specify the antigen or antibody you will need to do the assay you choose. Most of these questions have more than one right answer.

a. Detecting a large serum protein.
b. Detecting an antibody to a hapten.
c. Typing ABO antigens on erythrocytes.
d. Quantifying CD25 on lymphocytes.
e. Quantifying thyroid hormone levels in serum.
f. Identifying HIV gp41 on an electrophoresis gel.
g. Identifying B cells in the spleen under the light microscope.
h. Visualizing antigen-binding B cells under the electron microscope.
i. Separating B cells from T cells.
j. Separating IgG from IgE.
k. Detecting immune complexes in the serum.
l. Determining the isotype of your monoclonal antibody.

SOLUTION

a. Precipitation (Ouchterlony) is less sensitive but possible with a large protein. ELISA and RIA are more sensitive and work well with all soluble antigens. Purified antibody to the protein is needed, appropriately labeled for ELISA or RIA.
b. ELISA is the best assay; precipitation requires an antigen large enough to bind at least two antibody molecules and haptens are very small. You will need purified antibody to the hapten and enzyme-labeled anti-IgG.
c. For RBC antigens, agglutination or complement-mediated lysis with specific anti-A and anti-B antibodies both work well.
d. Flow cytometry uses fluorescent anti-CD25 to count the number of $CD25^+$ cells and determine the amount of CD25 on each cell.
e. ELISA with antibody to thyroid hormone.
f. Western blotting uses enzyme-labeled anti-gp41 to bind gp41 that has been separated by gel electrophoresis and blotted onto a nitrocellulose membrane.
g. Immunofluorescence, autoradiography, and immunohistochemistry with appropriately labeled anti–B-cell receptor (or antibodies to other B-cell markers) all will work.
h. Immunoelectron microscopy with an electron-dense label on the antigen.
i. From lowest to highest purity: nylon wool column (B cells bind, T cells pass through); affinity chromatography and paramagnetic beads with either anti-IgM for B cells or anti-CD3 for T cells; FACS using the same antibodies conjugated to fluorochromes.
j. Affinity chromatography using antibody specific for either IgG or IgE.
k. Complement-fixation assay will indicate the presence of immune complexes.
l. ELISA or Ouchterlony with isotype-specific antibody.

Chapter Test

True/False

1. The best technique for counting the number of T_C cells in a mouse lymph node would be flow cytometry with fluorescent anti-CD8 antibody.

2. Virus neutralization by IgA blocks infectious virus from entering host cells.
3. The most sensitive technique for detecting antibody is affinity chromatography.
4. T-cell activation can be measured by the effects of the T cells on other cells.
5. Paramagnetic beads coated with antibody to CD19 could be used to remove B cells from a suspension of spleen cells.

Multiple Choice

6. An antibody titer measures the
   a. Ability of antibody to neutralize a toxin.
   b. Amount of antibody to a specific antigen.
   c. Length of time since antigen contact.
   d. Number of antigen epitopes.
   e. Specificity of antibody for antigen.
7. A virus vaccine that can activate cytotoxic T cells *must* contain
   a. A high dose of virus particles.
   b. An adjuvant to stimulate T-cell division.
   c. Foreign MHC.
   d. Live virus.
   e. Virus peptides.
8. Antigens that are best able to activate helper T cells
   a. Are chemically complex.
   b. Are large.
   c. Are virus surface carbohydrates.
   d. Can be presented efficiently by MHC class II.
   e. Have repeating epitopes.
9. In an adoptive transfer experiment to study the interaction of APC, helper T cells, and B cells in antibody synthesis, the irradiated mouse serves as the
   a. Antigen donor.
   b. B-cell donor.
   c. MHC donor.
   d. T-cell donor.
   e. "Test tube" in which all cells can interact free of host cell immune function.
10. A knock-out mouse for T-cell receptor would be similar in phenotype (function) to a(n)
    a. Congenic mouse.
    b. Inbred mouse.
    c. Mouse with a T-cell hybridoma.
    d. Nude mouse.
    e. SCID mouse.

Short Answer

11. What is the role of adjuvants in immunization?
12. In general, how do we detect humoral immunity?

13. Why is FACS usually preceded by affinity or magnetic-bead enrichment?
14. You find a tube in your laboratory refrigerator marked "FITC-rabbit anti-mouse CD3" or "FITC-rabbit anti-mouse CD8"—the last character is blurred. You also have a TR-rabbit anti-mouse CD4. Fluorescein isothiocyanate (FITC) is a fluorochrome that glows green; TR (texas red) is a red-emitting fluorochrome. How can you determine the specificity of the unknown antibody?

Essay

15. You would like to make monoclonal antibody specific for a mouse B-cell leukemia cell line. You know the leukemia is a monoclonal tumor of B cells, but you do not know the antigen specificity of its membrane Ig. How could you make a monoclonal antibody that would specifically bind the tumor cell line?

Chapter Test: Answers

1. **T** 2. **T** 3. **F** 4. **T** 5. **T** 6. **b** 7. **d** 8. **d** 9. **e** 10. **d**

11. Adjuvants are molecules such as bacterial products and alum that increase immunogenicity by attracting and activating macrophages.
12. Humoral immunity is detected and quantified by measuring antibody production.
13. FACS is relatively slow; it is more efficient to use it to increase purity from more than 90% to 99% than from less than 50% to 99%, so other enrichment methods are used first.
14. Use the unknown antibody to stain mouse spleen cells along with the TR-anti-mouse CD4 and analyze the results by flow cytometry. Rabbit anti-mouse CD3 should stain all T cells; anti-mouse CD8 should stain only T cells that do not stain with anti-CD4, because the mature T cells in the spleen have either CD4 or CD8.
15. Because the leukemia probably resembles a normal B cell in most of its membrane antigens, inject a mouse of the same strain with the leukemia cell line (you do not want to generate anti-mouse MHC). Make hybridomas from spleen cells of the immunized mouse and myeloma cells. Select for hybridomas producing antibodies that bind the leukemia cell line but do not bind normal B cells from the same strain; they should be specific for the variable region of the B-cell receptor or for another leukemia-specific antigen.

Check Your Performance:

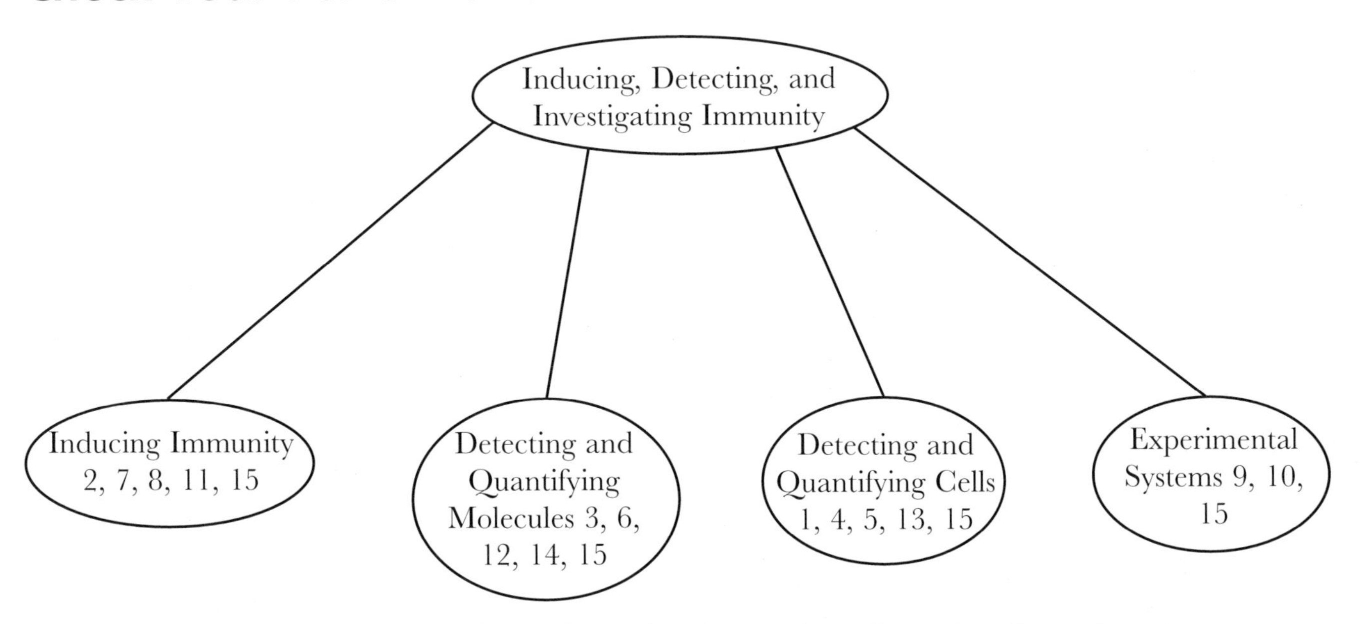

Check your understanding of this chapter by noting the number of questions for each topic you missed on the chapter test.

UNIT II: ANTIGEN RECOGNITION

Antibody Structure and Function

Antibodies are the effector molecules of humoral immunity. They bind antigen very specifically to neutralize its binding to host cells, stimulate phagocytosis, and activate the classical complement cascade.

ESSENTIAL BACKGROUND

- **Protein structure**
- **Antigen structure (Chapter 1)**
- **Affinity chromatography (Chapter 2)**

TOPIC 1: BASIC ANTIBODY STRUCTURE

KEY POINTS

✓ *Which structural features are shared by all antibody molecules?*

✓ *How does antibody structure correlate with its functions?*

✓ *What are the five immunoglobulin isotypes?*

Antibodies are plasma glycoproteins, also known as gamma globulins because of their electrophoretic mobility and immunoglobulins (Ig) because of their role in immunity. All antibodies share a basic structure (**Figure 3.1**). Each antibody "monomer" has a molecular weight of approximately 150 kDa and is composed of two identical polypeptide **heavy** (H) **chains** of about 440 amino acids and two identical **light** (L) **chains** of about 220 amino acids, covalently bonded via interchain disulfide linkages. H and L chains each contain intrachain disulfide bonds that stabilize their folding into **domains**, each about 110 amino acids long. Ig domains are a common feature of many soluble molecules and membrane-bound receptors of the immune system (the Ig superfamily). Amino acid sequences of the amino terminal H and L chain domains vary considerably and are responsible for the antigen-binding diversity of antibodies; these regions are the **variable regions** V_H and V_L. Within V_H and V_L there are **hypervariable regions** and less variable **framework regions**. Folding brings the hypervariable regions together to form the antigen-binding pockets; sites of closest contact between antibody and antigen are antibody **complementarity determining regions** (**CDR**). Amino acid sequences of the carboxyl L chain domain and three H chain domains have relatively limited variability; they are the **constant regions** (C_H and C_L).

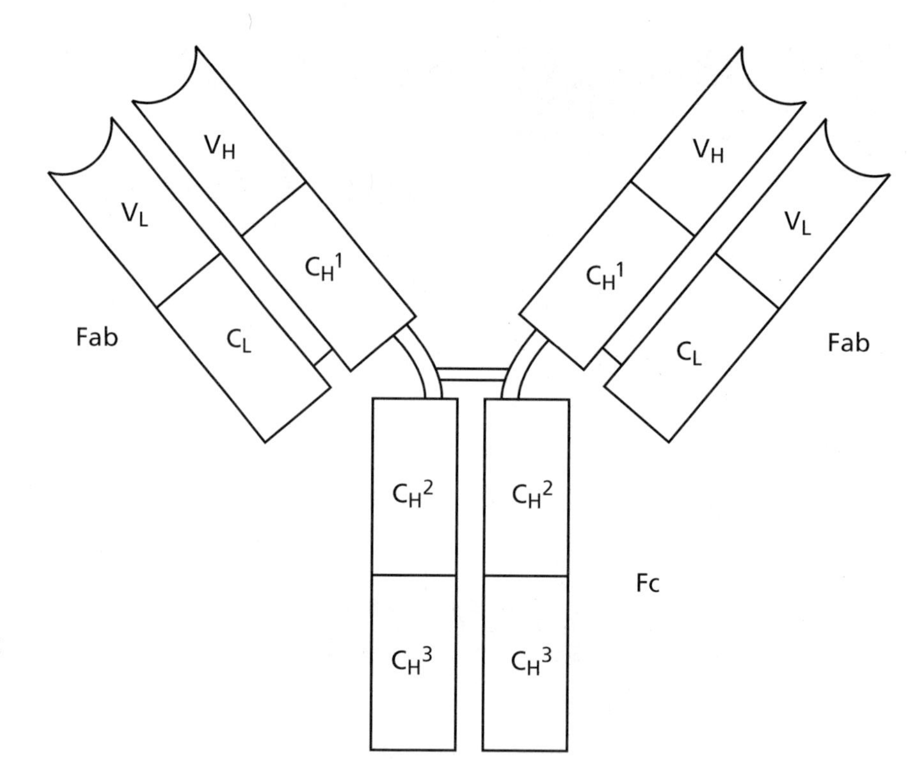

Figure 3.1 Immunoglobulin structure. Heavy chains determine the isotype; light chains are identical in each molecule but may be kappa or lambda.

Limited proteolytic digestion with papain cleaves Ig monomer into three fragments. Two identical amino terminal fragments containing one entire L chain and about half an H chain are the antigen-binding fragments (**Fab**). The third fragment, similar in size but containing the carboxyl terminal half of both H chains with their interchain disulfide bond, is the **Fc** fragment. Fc contains the carbohydrate residues and complement-binding and Fc receptor (FcR)-binding sites. A more extended region between H chain Fab and Fc is the **hinge**; it allows the two Fab regions to move independently to bind antigen. Limited pepsin digestion yields a single $(Fab')_2$ fragment containing both Fab pieces and the hinge region, including the H–H interchain disulfide bond; it is divalent for antigen binding. Pepsin digests the carboxyl halves of the H chains.

Sequence analysis of myeloma proteins, homogenous antibodies produced by plasma cell tumors, showed that all antibodies have one of two kinds of L chain, **kappa** or **lambda**. Five different H chain isotypes have been found: **alpha**, **gamma**, **delta**, **epsilon**, and **mu**. Antibody isotypes (classes) are named IgA, IgG, IgD, IgE, and IgM to correspond to their H chain types, which influence the effector functions of the antibody molecules.

Topic Test 1: Basic Antibody Structure

True/False

1. A molecule of IgG and a molecule of IgM can both have kappa light chains.
2. Regions of antibody H and L chains that contribute *most* to antigen binding are the CDR

Multiple Choice

3. An antibody Fab contains
   a. Complementarity determining regions.

b. H and L chain variable regions.
c. One antigen binding region.
d. One H–L interchain disulfide bond.
e. All of the above

4. Myeloma proteins are
   a. Abnormally formed antibodies secreted from cancerous plasma cells.
   b. Cancerous plasma cells that divide without requiring antigen activation.
   c. Cell lines that secrete specific antibodies for a short time and then die.
   d. Homogenous antibody molecules secreted by plasma cell tumors.
   e. Protein signaling molecules that make a plasma cell become a multiple myeloma.

Short Answer

5. What does the "prime" (′) in $(Fab')_2$ signify?
6. Draw an immunoglobulin monomer; label the H and L chains, Fab and Fc regions, variable and constant regions, and hinge region and indicate the location of the interchain disulfide bonds.

Topic Test 1: Answers

1. **True.** All light chains in a single Ig molecule are identical and either kappa or lambda; kappa and lambda light chains are found in all antibody isotypes.
2. **True.** CDR contain the amino acids that have closest contact with antigen.
3. **e.** Fab contains an entire L chain and the amino terminal half of one H chain.
4. **d.** Multiple myeloma is a plasma cell tumor; the antibodies secreted by the plasma cells are normal. Because they come from a tumor arising from a single cell, myeloma proteins have identical H and L chain sequences.
5. The prime indicates the additional part of the H chains containing the H–H disulfide bonds not found in single Fab fragments.
6. Review Figure 3.1.

TOPIC 2: ANTIBODY INTERACTION WITH ANTIGEN

KEY POINTS

✓ *What physical forces hold antigens and antibodies together?*

✓ *What is antibody cross-reactivity?*

✓ *What is the difference between an antibody's affinity and avidity for antigen?*

Antibodies bind antigens using a variety of noncovalent forces: hydrogen bonds, ionic bonds, hydrophobic interactions, and Van der Waals interactions. The antigen-binding cleft of a typical antibody can accommodate approximately four to seven amino acids or sugar residues. Contact between large antigens and antibodies is probably more extended. The specificity of antigen binding resembles that of enzyme-substrate binding.

Viral capsid proteins or bacterial cell wall components usually have several epitopes, each one eliciting a specific antibody response in relation to its immunogenicity. Antibodies can distinguish antigenic differences (serotypes) between members of the same species. Some epitopes are shared by different antigens, so that antibody made to one also binds the other. Cross-reactive antibodies are one mechanism by which autoimmunity is induced; in certain individuals, antibodies made to *Streptococcus pyogenes* bind a cross-reactive antigen on their heart valves and may cause rheumatic heart disease.

The strength of the interaction between a single antibody paratope and its specific antigen epitope is called **binding affinity**. The higher the affinity, the tighter the association between antigen and antibody. Antibody affinity generally increases with repeated exposure to antigen because B cells with higher affinity antigen receptors produce larger clones of antibody-secreting plasma cells. The affinity constant K_a is the ratio between the rate constants for binding and dissociation of antibody and antigen. Typical affinities for IgG antibodies are 10^5 to 10^9 liter/mole. Antibody affinity is measured by equilibrium dialysis. The relationship between bound and free antigen and antibody affinity is expressed by the Scatchard equation, $r/c = Kn - Kr$, where r is the ratio of bound antigen concentration to total antibody concentration, c is the concentration of free antigen, K is the affinity constant, and n is the number of binding sites per antibody molecule. If all the antibodies have the same affinity, a plot of r/c versus r yields a straight line with a slope of $-K$ and an r intercept approaching n. Heterogeneous antibody yields a curved line; the average affinity can be determined by the slope of the curve when half the binding sites are full ($r = 1$).

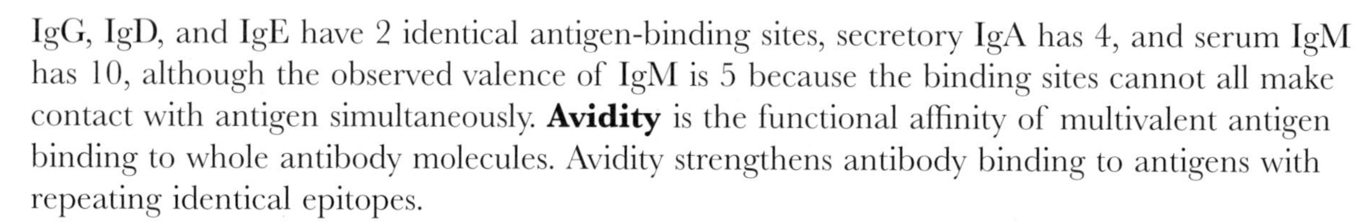

IgG, IgD, and IgE have 2 identical antigen-binding sites, secretory IgA has 4, and serum IgM has 10, although the observed valence of IgM is 5 because the binding sites cannot all make contact with antigen simultaneously. **Avidity** is the functional affinity of multivalent antigen binding to whole antibody molecules. Avidity strengthens antibody binding to antigens with repeating identical epitopes.

Topic Test 2: Antibody Interaction with Antigen

True/False

1. Contact between antigen epitope and antibody paratope involves about two to four antigen amino acid residues.
2. IgM is the isotype with the highest valence for antigen.

Multiple Choice

3. Avidity
   a. Is a pathogenic agent, causing a very serious disease.
   b. Is the interaction between antibody and antigen that results in visible clumping.
   c. Occurs when the ratio of antibody to antigen is optimal.
   d. Refers to the strength of interactions between a multivalent antibody and antigen.
   e. Results in a loss of antibody reactivity.
4. A colleague sends you an antibody to influenza virus hemagglutinin. You perform equilibrium dialysis on the antibody to measure its affinity. Plotting r/c versus r gives you a

curved line with K = 10^9 liters/mole and an r intercept of 4. From these results, you conclude that the antibody is probably

a. A cross-reactive antibody.
b. A monoclonal anti-influenza hemagglutinin.
c. A polyclonal IgG antibody.
d. IgA anti-influenza hemagglutinin.
e. Not specific for influenza hemagglutinin.

Short Answer

5. What is antibody cross-reactivity?
6. Describe in words the concept of affinity.

Topic Test 2: Answers

1. **False.** The binding pocket on the antibody will hold about four to seven amino acids, but contact with antigen is usually over a wider area.
2. **True.** Serum IgM is a pentamer; it can usually bind at least five epitopes.
3. **d.** The binding strength of intact antibodies is due to multivalent binding to repeating antigen epitopes, common on bacterial and viral surfaces.
4. **d.** A curved Scatchard plot means the preparation contains antibodies of several different affinities, so it must be polyclonal. Because the r intercept is 4, the antibody must be an IgA dimer that has four binding sites, not an IgG monomer that has two. The affinity is high, so the antibody is specific for the antigen.
5. Cross-reactive antibodies bind structurally similar epitopes on another antigen than the one to which they were produced.
6. The affinity constant expresses the likelihood that antigen and antibody will be bound to each other instead of free. As K_a increases, more antibody molecules will contain antigen. More complementarity between antigen and antibody increases the K_a.

TOPIC 3: FUNCTIONAL CLASSES OF ANTIBODY

Key Points

✓ *What are the biological functions of the five antibody isotypes?*

✓ *What benefits are associated with having different isotypes?*

✓ *What are immunoglobulin allotypes and idiotypes?*

Biological functions of antibody are independent of antigen specificity but dependent on isotype. Antibody is produced on polyribosomes, where signal recognition protein attached to leader sequence sends the protein into the endoplasmic reticulum, where H and L chains assemble and carbohydrate and J chains are added. The vesicle containing antibody moves to the plasma membrane and exocytosis releases secreted antibody from the plasma cell. Membrane-bound Ig has an additional transmembrane sequence on its carboxyl terminal C_H region.

IgM is the first B-cell receptor made during B-cell development and the first antibody secreted during an immune response. Membrane IgM is a four-chain μ_2L_2 monomer; serum IgM is a pentamer of five four-chain monomers held together by interchain disulfide bridges in C_H3 and C_H4 plus an extra peptide called **J chain**. The pentameric nature of IgM makes it the best complement-fixing antibody but too large to easily leave the circulation. Low levels of IgM are present in mucosal secretions.

IgG, the predominant serum antibody, has the longest half-life. Four subclasses of IgG each have somewhat different biological functions. IgG is made later in a primary response than IgM but is produced more rapidly in a memory response. IgG crosses the placenta to transfer maternal immunity to the fetus and enters the tissues to neutralize virus and toxin binding to host cells. Two molecules of IgG are required to activate complement. IgG–antigen complexes bound to FcR stimulate phagocytosis.

IgA is present in serum and predominates in mucosal secretions: breast milk, saliva, tears, and respiratory, digestive, and genital mucus. Secretory IgA provides a first-line defense where pathogens enter the body; more IgA is made than any other isotype. Serum IgA is usually monomeric, although dimers, trimers, and tetramers are present. **Secretory IgA** is dimeric or tetrameric and contains one J chain and one additional chain called **secretory component**, which protects it from degradation. The plasma cell makes IgA and J chain and then assembles and secretes polymeric IgA. IgA then binds to a poly-Ig receptor on mucosal epithelial cells, crosses the cytoplasm, and exits on the luminal side with part of the poly Ig receptor as secretory component.

IgE is responsible for the symptoms of allergy because it binds very efficiently to mast cell FcεR. Antigen cross-linking of IgE on FcεR signals the mast cell to release histamine. IgE also helps eosinophils destroy helminth parasites. IgD, with IgM, is the B-cell receptor for antigen. Its presence on the B-cell membrane signals that the B cell is mature and ready to respond to antigen. IgD is present in serum in low amounts; no effector functions have been identified for secreted IgD.

Immunoglobulins can be used as antigens to generate antibodies that distinguish several epitopes: isotypes, allotypes, and idiotypes. All members of a species share **isotypes**; anti-isotype sera differentiate epitopes in C_H and C_L. Within a species and isotype, amino acid sequence variations in C_H and C_L differentiate Ig **allotypes**. Allotypic epitopes have been identified in all human subclasses of gamma, alpha$_2$, and kappa chains and are called Gm, Am, and Km. Each individual has antibodies with one (homozygous) or two (heterozygous) allotypes; an individual B cell produces antibody of a single allotype (allotypic exclusion). Idiotypic epitopes are due to sequence differences within V_H and V_L, so each individual makes antibodies with many **idiotypes**. Monoclonal antibody molecules share the same idiotype; polyclonal antibodies, even those made against the same epitope, may have different idiotypes. Anti-idiotype and antigen usually compete for the antigen-binding region of an Ig.

Topic Test 3: Functional Classes of Antibody

True/False

1. Pentameric IgM is the antigen receptor on mature B cells.
2. Antigen cross-linking of IgE bound to mast cells causes release of worm parasites.

Multiple Choice

3. Human serum IgA is isolated and injected into a rabbit. The resulting rabbit anti-IgA antibodies will react against all of the following *except* human
   a. Alpha chain.
   b. IgG.
   c. J chain.
   d. Kappa chain.
   e. Secretory component.

4. You have purified some Fab from an IgG myeloma protein. Under appropriate conditions, you could use this Fab to generate antibodies to
   a. Both kappa and lambda chains.
   b. Gamma chain hinge region.
   c. J chain.
   d. The gamma chain isotypic determinants.
   e. The idiotype of this myeloma.

Short Answer

5. How do IgM and IgE monomers differ from the IgG monomer you drew earlier?

6. How might antibody allotypes be used in paternity testing?

Topic Test 3: Answers

1. **False.** Membrane IgM is a monomer; pentameric IgM is secreted.

2. **False.** Antigen-IgE binding on mast cells triggers histamine release.

3. **e.** Antiserum to IgA contains antibodies to alpha, kappa, lambda, and J chains, because serum IgA is polyclonal and occasionally polymeric. Serum IgA does not contain secretory piece.

4. **e.** Fab does not contain the H chain isotypic or hinge determinants. Because this is a monoclonal antibody (mAb), it will have either a kappa or lambda chain, not both.

5. Mu and epsilon chains have four C_H regions, compared with three in the gamma chain, and do not have an extended hinge region

6. Allotypes are sequence variations in C_H and C_L; a child's antibodies would share allotypes with each parent. An allotype in the child's Ig that is not present in the mother's or possible father's Ig would rule out paternity; finding only the allotypes of mother and possible father would not prove paternity, because many people have the same allotypes. Today's paternity tests match much more genetic material.

TOPIC 4: DESIGNER ANTIBODIES

KEY POINTS

✓ *What is a monoclonal antibody and how is one obtained?*

✓ *When are monoclonal antibodies superior to polyclonal antibodies?*

✓ *What are the advantages of humanized antibodies, chimeric antibodies, immunotoxins, and heteroconjugates over standard monoclonal antibodies?*

Myeloma proteins are examples of mAb. They are homogenous antibodies produced by a clone of plasma cells originating from a single cell that mutated to divide continuously. Myelomas arise spontaneously in humans and animals; the antigen specificity is usually unknown. mAb of a desired specificity can be produced in the laboratory by making hybridomas (Chapter 2). mAb are superior to polyclonal antibodies for many research, diagnostic, and therapeutic purposes because of their defined epitope specificity. They are coupled with fluorescent dyes to label cells in vitro, with radioactive probes for imaging tumors in vivo, and with toxins to kill tumor cells.

Immunologists use genetic engineering to modify mAb. Mouse mAb injected repeatedly into humans induce a human anti-mouse antibody (**HAMA**) immune response that reduces their effectiveness. Fab or $(Fab')_2$ fragments are less immunogenic but have a shorter serum half-life. **Chimeric antibodies** are produced by cells engineered with mouse V_H and V_L gene segments of the desired specificity spliced to human C_H and C_L gene segments. In some cases, mAb are further **humanized** by splicing mouse CDR with human framework and C region gene segments. **Fv antibody** is engineered to contain only V_H and V_L transcribed as a single polypeptide. Toxin sequences replace the Fc region of antibody in engineered **immunotoxins. Heteroconjugates** are made with H–L pairs from different antibodies. If an H–L specific for tumor antigen were disulfide bonded with an H–L specific for cytotoxic T lymphocyte (T_C) CD8, the heteroconjugate would cross-link cytotoxic T cells with tumor cells and promote tumor cell lysis.

Topic Test 4: Designer Antibodies

True/False

1. A humanized antibody has mouse CDR and human framework and C regions.
2. Monoclonal antibodies are better than polyclonal antibodies at activating complement to kill tumor cells.

Multiple Choice

3. You have developed a mouse mAb to a melanoma skin tumor antigen and want to inject it into melanoma patients to eliminate tumor cells that have metastasized. To efficiently kill the tumor cells, you would probably
   a. Create a heteroconjugate between your antibody and an mAb against CD4 to link T_C with tumor cells.
   b. Inject the mouse antibody into a human to humanize it.
   c. Make an immunotoxin by linking your antibody with a bacterial toxin.
   d. Stabilize your antibody by splicing mouse Fc with human Fab.
   e. All of the above strategies would increase the effectiveness of your antibody.
4. Compared with mAb from a myeloma, mAb from a hybridoma have a
   a. Higher affinity for antigen.
   b. Known antigen specificity.

c. Single antibody allotype.
d. Single antibody idiotype.
e. Single antibody isotype.

Short Answer

5. What is a hybridoma?
6. Why is it advantageous for the body to make a polyclonal response rather than a monoclonal response to a pathogen?

Topic Test 4: Answers

1. **True.** A humanized antibody is human except for its CDR.
2. **False.** Complement activation depends on the antibody isotype.
3. **c.** Anti-CD8, not anti-CD4, binds T_C. Injecting mouse antibody into a human would generate a HAMA response; to reduce the HAMA response, use mouse CDR with human framework and C regions. Linking your mAb to toxin would deliver the toxin to the target.
4. **b.** Answers c, d, and e are true for monoclonals from both sources. The affinity of each monoclonal is different, but some are high and some are low.
5. A hybridoma is a cell made by fusing a plasma cell and a myeloma cell.
6. Diversity of response increases its chance of success. Antibody binding certain epitopes may block pathogen binding to host cells, whereas other antibodies neutralize toxins. Some pathogens change their antigens, evading a response to a single antigen. The high avidity of IgM activates complement, but IgG enters the tissues more efficiently and IgA can be secreted to bind antigen before it enters the body.

CLINICAL CORRELATION: MULTIPLE MYELOMA

Multiple myeloma is a rapidly dividing plasma cell tumor secreting a monoclonal Ig. Patients with myeloma have elevated serum antibody levels with restricted electrophoretic mobility and depressed levels of normal antibodies. Hematopoiesis, bone calcification, and immunity are depressed, resulting in anemia, bone fractures, and susceptibility to exogenous pathogens like *Streptococcus pneumoniae* and *Staphylococcus aureus,* which are most efficiently eliminated by neutrophils and opsonizing IgG antibodies. Excess L chains (**Bence Jones proteins**) are excreted in the urine. Multiple myeloma is treated with transfusions as needed to alleviate anemia and antitumor therapy. Because high serum levels of antibody increase Ig turnover, passive gamma globulin therapy is not effective in multiple myeloma.

DEMONSTRATION PROBLEM

To generate IgG antibody to human serum albumin (HSA), you inject a rabbit on several occasions with pure HSA and collect the rabbit antiserum. An ELISA with the antiserum and several purified proteins gives the following results: HSA, 100 binding units; human IgG, 0 units; rabbit serum albumin (RSA), 2 units; and chimpanzee serum albumin (CSA), 98 units. By amino acid sequencing, HSA and CSA are about 90% similar and HSA and RSA are about 45% similar. How would you purify the anti-HSA from the rabbit serum and determine its isotype, and what can you deduce about the cross-reactivity and immunogenicity of these four serum molecules from the antibody binding data?

SOLUTION

Antibody can be separated from other serum proteins by differential solubility in ammonium sulfate or by affinity chromatography on a column of anti-rabbit IgG. To purify HSA-specific antibody, run an affinity column with HSA on the beads; specific antibody will bind and can be eluted with dilute acid or chaotropic agents, which weaken the noncovalent bonds between antigen and antibody. Antibody isotype can be determined by reacting it with anti-isotype–specific antisera in immunoblot, Ouchterlony, ELISA, or radioimmunoassay (Chapter 2). The binding results suggest that the anti-HSA is highly cross-reactive with chimpanzee albumin, which you would have predicted from their sequence similarity. Even though RSA shares about 45% of its sequence with HSA, it does not bind significantly to anti-HSA, probably because the rabbit did not respond to epitopes that resembled "self." IgG does not seem to share epitopes with HSA.

Chapter Test

True/False

1. All animals in a species express the same immunoglobulin isotypic determinants.
2. Mouse anti-human IgM binds human B cells.
3. The binding strength of the multiple interactions between a multivalent antibody and antigen is its affinity.
4. Rabbit anti-mouse IgG binds mouse kappa chain.
5. Humanized antibodies are less immunogenic in humans than mouse antibodies.

Multiple Choice

6. IgD
   a. Binds through its Fc to mast cells and basophils.
   b. Is the first antibody to appear during an immune response.
   c. Is an antigen receptor on B cells.
   d. Is a pentamer.
   e. None of the above
7. Which of the following changes to a secretory IgA antibody molecule would definitely *decrease* its avidity?

a. Increase noncovalent antigen–antibody interactions in the CDR.
b. Remove the secretory component.
c. Replace the Fc portion of the alpha chains with the Fc portion of mu chains.
d. Replace V_H and V_L framework regions with those from a different antibody.
e. Use limited enzyme digestion to make Fab fragments.

8. Allotypic determinants are
   a. Constant region determinants that distinguish each Ig class and subclass within a species.
   b. Expressed only from the paternal chromosome.
   c. Generated by the conformation of antigen-specific V_H and V_L sequences.
   d. Not immunogenic in individuals who do not have that allotype.
   e. Amino acid differences encoded by different alleles for the same H or L chain locus.

9. The regions of the antibody molecule that contribute *most* to the affinity of the antibody for antigen are the
   a. CDR.
   b. Fab regions.
   c. Fc regions.
   d. Framework regions.
   e. Hinge regions.

10. mAbs are ___________ than polyclonal antibodies.
    a. Higher affinity
    b. Less cross-reactive
    c. Less heterogeneous
    d. More cytotoxic
    e. More humanized

Short Answer

11. What antibody effector functions involve FcR?
12. What is J chain and which immunoglobulins contain it?
13. How could you physically separate free immunoglobulin H and L chains? Could you use their antigenic properties or their antigen-binding properties to separate them?
14. How would you make an anti-idiotype?

Essay

15. Polio virus infects intestinal epithelial cells, which replicate more polio virus; it can spread to cause paralysis by infecting motor neurons. The Salk injected killed polio vaccine stimulates production of mainly IgG antibody to the polio virus capsid protein. The Sabin oral attenuated polio vaccine causes a mild infection of intestinal epithelial cells and stimulates production of secretory IgA antibody to the capsid protein. Compare and contrast the biological activities of the two antibody isotypes that protect against polio infection.

Chapter Test: Answers

1. **T** 2. **T** 3. **F** 4. **T** 5. **T** 6. **c** 7. **e** 8. **e** 9. **a** 10. **c**

11. FcR are membrane molecules that bind Ig Fc regions. Antigen–antibody complex binding to phagocyte FcγR promotes antigen uptake; binding to FcεR on mast cells promotes histamine release.
12. J chain is a polypeptide chain found on polymeric IgM and IgA. It binds poly Ig receptor on epithelial cells.
13. Reduce interchain disulfide bonds and separate the chains by molecular weight or by affinity chromatography with anti-isotype antibodies (isotypes are antigenic properties). Isolated chains usually have low affinity for antigen.
14. Inject an mAb of the desired idiotype into an individual of the same species and allotype so that only idiotype is foreign.
15. Serum IgG enters intercellular tissue spaces to neutralize polio virus, activates complement to attract phagocytes, and opsonizes the virus for elimination. Secretory IgA in the intestinal secretions blocks polio virus binding to intestinal epithelial cells; IgA in the serum and tissues neutralizes virus but does not activate complement.

Check Your Performance:

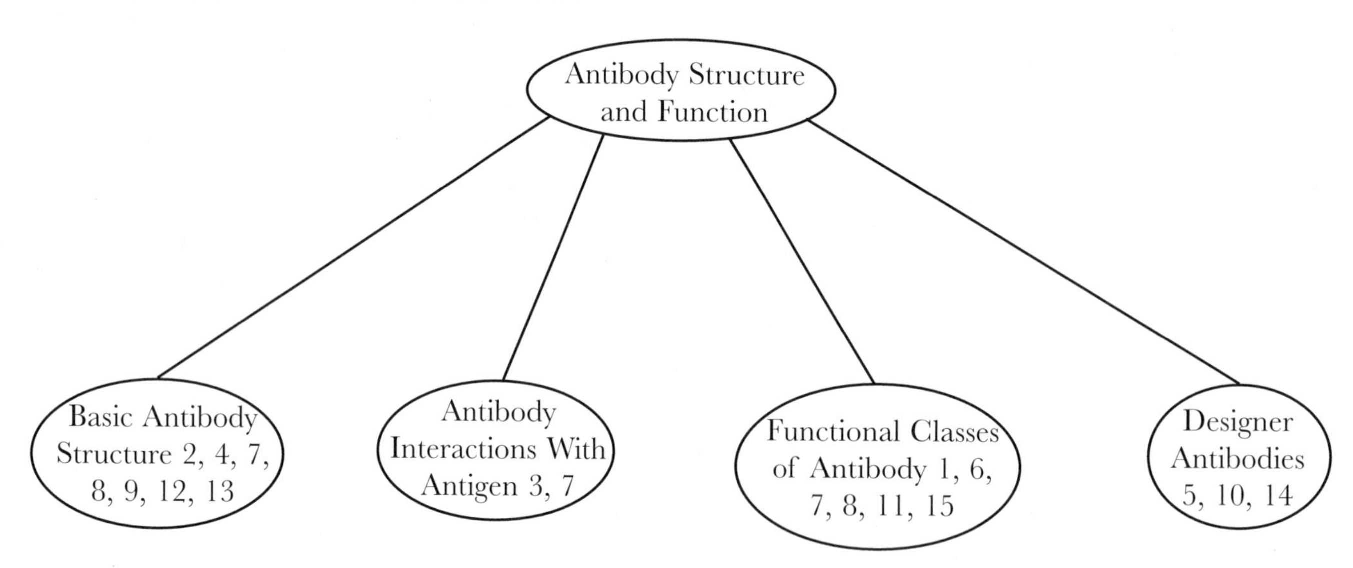

Check your understanding of this chapter by noting the number of questions for each topic you missed on the chapter test.

Antibody Gene Organization and Expression

The immune system is remarkable for its ability to respond to many antigens. Unusual properties of antibody diversity include the presence of variable and constant regions on the same polypeptide chain and identical V regions used with different C regions. The somatic recombination process for generating antibody and T-cell receptor diversity is unique among mammalian systems.

ESSENTIAL BACKGROUND

- **Gene structure, transcription, and translation**
- **Clonal selection (Chapter 1)**
- **Antibody structure and function (Chapter 3)**

TOPIC 1: ORGANIZATION OF IMMUNOGLOBULIN GENES

KEY POINTS

✓ *How are light chain genes organized?*

✓ *How are heavy chain genes organized?*

✓ *Which genes encode each segment of the antibody molecule?*

Antibodies must have enough antigen-binding diversity to recognize every possible pathogen (many V regions) while maintaining the biological effectiveness of their C regions. **Somatic recombination** allows many V regions to be used with a few C regions. Gene segments encoding immunoglobulin (Ig) H, kappa, and lambda chains are on three different chromosomes. During B-cell development, **recombinases** remove introns and some exons from the DNA and splice segments into functional Ig genes.

Ig gene segments in mammals are arranged in groups of variable (V), diversity (D), joining (J), and constant (C) exons. Vκ encodes the first two complementarity determining regions (CDR) and three framework regions (FR) of Vκ, plus a few residues of CDR3. Jκ encodes the remainder of CDR3 and the fourth FR. Cκ encodes the complete C region of the kappa light chain. DNA encoding human kappa chain includes approximately 40 functional Vκ segments, five Jκ segments, and one Cκ gene segment and some gene segments that contain premature stop codons (pseudogenes, psi). Human lambda chain DNA contains approximately 30 functional

Vλ segments, four functional Jλ , and four functional Cλ. DNA for human H chain includes approximately 65 functional V_H segments, 30 D_H segments, and six J_H segments. The first two CDR and three FR of H chain variable region are encoded by V_H. CDR3 is encoded by a few nucleotides of V_H, all of D_H, and part of J_H, whereas FR4 is encoded by the remainder of the J_H gene segment. There are also individual gene segments for each heavy chain domain and membrane region of each isotype, arranged in the order shown in **Figure 4.1**.

Topic Test 1: Organization of Immunoglobulin Genes

True/False

1. Ig genes for H, kappa, and lambda chains are many kilobases apart on the same chromosome.

2. CDR3 for both H and L chains are encoded by more than one gene segment.

Multiple Choice

3. Genes for Igs are unlike other human genes in that
   a. Each polypeptide chain is encoded by several exons.
   b. Ig genes are composed of introns and exons.
   c. Somatic recombination occurs before mRNA is transcribed.
   d. There is less Ig genetic material in mature B cells than in other somatic cells.
   e. Both c and d

4. The gene segments needed to encode the variable region of a kappa chain are
   a. One Jκ plus one Dk.
   b. One Jκ plus one Cκ.
   c. One Vκ plus one Dk.
   d. One Vκ plus one Jκ.
   e. One Vκ plus one Jκ plus one Dk.

Short Answer

5. How many gene segments does it take to make a heavy chain variable domain?

6. Approximately how many possible human antibody kappa variable regions could be made by randomly combining any Vκ with any Jκ gene segment? Perform the same calculations for the human H chain variable region.

Topic Test 1: Answers

1. **False.** The gene segments for each antibody chain are on a different chromosome.

2. **True.** CDR3 for L chain is encoded partly by V_L and partly by J_L. CDR3 for H chain is encoded by V_H, D_H, and J_H.

3. **e.** Somatic recombination of Ig genes involves splicing out introns and exons to bring V, D, and J segments together. Because only B cells rearrange their Ig genes, they actually have less DNA for Ig than other somatic cells.

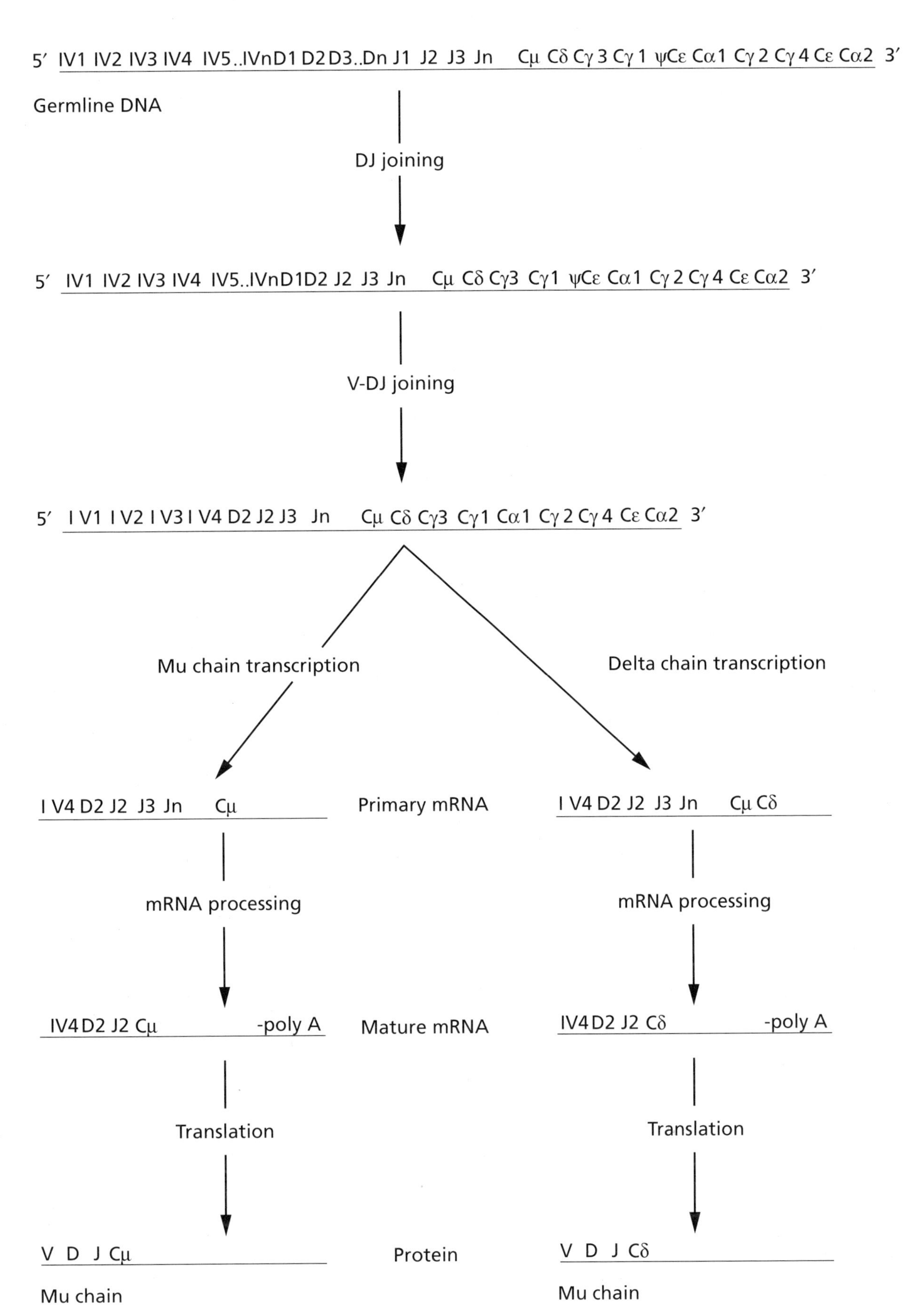

Figure 4.1 Somatic recombination of human heavy chain gene segments.

4. **d.** One Vκ plus one Jκ gene segment encode an entire kappa variable region.
5. Three: a V_H domain is encoded by a V_H plus a D_H plus a J_H gene segment.
6. A total of 200 possible human antibody kappa variable region chains could be made by randomly combining 40 Vκ with 5 Jκ gene segments. The same calculations for human H chain variable region gives $65 \times 30 \times 6 = 11{,}700$ possible V_H chain sequences.

TOPIC 2: SOMATIC RECOMBINATION

KEY POINTS

✓ *How are Ig gene segments rearranged into functional H and L chain genes?*

✓ *What information ensures that gene segments are rearranged appropriately?*

✓ *How does Ig gene rearrangement result in antigen-binding diversity?*

Somatic recombination occurs before antigen contact. One D_H and one J_H are randomly spliced with the removal of all intervening DNA (**D–J joining**; Figure 4.1). Next, a random V_H segment is spliced to rearranged DJ_H. VDJ_H is not spliced to C_H; the intervening sequences between VDJ_H and C_H are transcribed and then spliced out of the primary mRNA before mature message is translated into H chain. L encodes a leader (signal) sequence that is used to move Ig into the endoplasmic reticulum; it is not present in Ig protein. L chain gene segment recombination occurs with Vκ and Jκ joining; VJκ segments are transcribed with Cκ and intervening DNA. RNA splicing removes excess base-pairs to allow translation of a complete kappa chain. If kappa chain genes are not successfully rearranged, lambda chain gene segments are recombined. Each B cell makes antibody of a single antigen specificity and a single allotype. Both characteristics result from that B cell productively rearranging only one L and one H chain gene; **productive rearrangement** of one allele blocks the rearrangement of the other. If a B cell does not successfully rearrange both H and L chain genes, it dies.

V(D)J recombinase expressed in developing B cells recognizes **recombination signal sequences** of nine (nonamer) and seven (heptamer) base-pairs flanking each Ig gene segment. The spacers between heptamer and nonamer sequences are either 12 or 23 nucleotides long, signaling which gene segments may be joined. DNA between gene segments is usually looped out and removed during $V–J_L$ or $V–D–J_H$ joining, but in some cases it is inverted and retained. Splicing between gene segments is imprecise, sometimes resulting in nonproductive rearrangements in which frame shift mutations yield stop codons downstream. Products of at least two **recombination-activating genes**, *RAG-1* and *RAG-2*, plus enzymes to ligate the DNA and terminal deoxynucleotidyl transferase (TdT) are required for somatic recombination.

Combinatorial diversity is generated by random formation of many different VJ_L and VDJ_H combinations and is increased by the ability of most V_H regions to pair with most V_L regions to bind antigen. Random pairing of 320 different V_L with 11,700 different V_H results in almost 4×10^6 different possible antibody specificities. Actual combinatorial diversity is probably less, because some gene segments rearrange more than others and some V_HV_L pairings bind antigen better than others. **Junctional diversity** results from imprecise joining of gene segments and addition of nucleotides to the DNA sequence at splice sites; TdT adds up to 15 nucleotides to the DNA sequence of human V_H and J_H regions. Junctional diversity affects predominantly CDR3; it is significant but difficult to quantify because it also results in many nonproductive rearrangements. One additional mechanism for generating Ig diversity operates during antigen-dependent B-cell proliferation: **somatic hypermutation** of V region genes and selection for mutant antibodies with higher affinity for antigen (affinity maturation).

Topic Test 2: Somatic Recombination

True/False

1. During B-cell development, a V(D)J ligase removes introns and some exons from the DNA to produce functional Ig genes.
2. Primary mRNA for H chain encodes one V_H, one D_H, and multiple J_H segments.

Multiple Choice

3. Which does *not* contribute to Ig antigen-binding diversity?
   a. Any L chain can combine with any H chain to form a functional antibody.
   b. Any Vκ can be joined to any Jκ to encode the light chain V region.
   c. Many C_H genes are present in the germline DNA.
   d. Random numbers of N nucleotides can be added during somatic recombination.
   e. VJ_L and VDJ_H joining is imprecise.
4. The proper joining of one VJ_L and one VDJ_H is regulated by the
   a. Heptamer and nonamer sequences.
   b. Leader sequences.
   c. P-nucleotide addition sites.
   d. 12 and 23 nucleotide spacers between heptamer and nonamer sequences.
   e. TdT binding site for DNA.

Short Answer

5. Why does junctional diversity affect primarily CDR3?
6. A B cell that has productively rearranged its H chain genes does not productively rearrange either kappa chain gene but does go on to successfully produce lambda chain. Will this affect the antigen-binding specificity of the cell?

Topic Test 2: Answers

1. **False.** *RAG-1* and *RAG-2* genes encode a V(D)J recombinase expressed only in lymphocytes.
2. **True,** unless the most 3′ J_H segment is spliced to D_H. DNA between the spliced J_H and C_H is removed in RNA processing.
3. **c.** C_H affects antibody effector function, not antigen-binding specificity.
4. **d.** Recombination only occurs between recombination signal sequences with unlike spacers to ensure that V_L–V_L and V_H–J_H joining do not occur.
5. CDR1 and CDR2 are completely encoded within the Vκ, Vλ , and V_H gene segments and their amino acid sequences are not affected by somatic recombination.

6. Cκ or Cλ do not directly affect the antigen-binding specificity of the cell. Because Vκ and Jκ sequences differ from Vλ and Jλ, the antigen binding specificity will be affected.

TOPIC 3: B-CELL DEVELOPMENT

KEY POINTS

✓ *What developmental landmarks characterize B-cell development?*

✓ *What is the structure of the mature B-cell antigen receptor?*

✓ *What regulates antibody gene expression in B cells?*

In the bone marrow, cytokines are made that induce TdT and recombinase synthesis in $CD34^+$ lymphoid progenitors. The cells undergo D–J_H joining to become **pro-B cells** and also express CD45 (B220). Human B cells can undergo D–J_H joining on both chromosomes and D_H can be read in any reading frame, so all DJ_H rearrangements are productive. Pro-B cells become **pre-B cells** when they undergo V–DJ_H joining on one chromosome and express membrane mu chains with **surrogate L chains**, composed of germline VpreB and λ5. Signal transduction molecules Ig alpha and Ig beta are also part of the **pre-B receptor complex**, which allows the cell to receive a signal that halts H chain rearrangement and signals the pre-B cell to divide. The cytoplasmic tails of Ig H chains are short; those of Ig alpha/Ig beta have **immunoreceptor tyrosine activation motifs** that become phosphorylated in response to ligand binding and initiate a cytoplasmic signaling cascade. If H chain recombination fails on one chromosome, it occurs on the other; if nonproductive rearrangements occur on both chromosomes, the cell dies.

After proliferation, pre-B cells undergo V–Jκ joining on one chromosome. If rearrangement is productive, kappa chain is made and the cell becomes a mature B-cell expressing membrane IgM(κ) and IgD(κ) B-cell antigen receptor (BCR) complexed with Ig alpha/Ig beta. If kappa genes are not successfully rearranged on either chromosome after several attempts, lambda genes are rearranged; success leads to production of IgM(λ) and IgD(λ) BCR. If neither kappa nor lambda are productively rearranged, the cell dies in the bone marrow; only a minority of or a few human pre-B cells fail to become mature B cells. Before leaving the marrow or very shortly after entering the periphery, mature B cells undergo **negative selection**. If BCR binds multivalent self-antigens, the B cell dies (**clonal deletion**). If the B cell encounters soluble self-antigen, it becomes **anergic** (unreactive). Some self-reactive B cells perform **receptor editing** by rearranging another L chain or replacing V_H. If self-antigen binding does not occur, the mature B cell has a half-life of 3 to 8 weeks if it enters the lymphoid follicles and 3 days if it does not.

Ig gene expression is regulated by promoter, enhancer, and silencer DNA sequences. Somatic recombination of Ig genes increases the efficiency of Ig gene transcription by bringing promoter and enhancer regions closer together in the DNA. DNA binding proteins produced during development and after antigen and cytokine signaling also influence Ig gene expression.

Topic Test 3: B-Cell Development

True/False

1. Pre-B cells must receive a signal from specific antigen binding to pre-B BCR before they can proceed to the next stage in development.
2. Membrane Ig is always expressed with Ig alpha/Ig beta.

Multiple Choice

3. B-cell differentiation begins with the expression of
   a. DJCμ + surrogate light chain.
   b. Membrane IgD.
   c. Membrane IgM.
   d. VJCμ + surrogate light chain.
   e. *RAG-1*, *RAG-2*, and TdT.

4. Once H chain genes have been productively rearranged and expressed on pre-B cell membranes, the next event to occur in the cell is
   a. Proliferation of pre-B cells.
   b. Rearrangement of kappa chain gene segments.
   c. Receptor editing.
   d. V_H to DJ_H joining of other H chain alleles.
   e. VJλ joining.

Short Answer

5. Which events in B-cell development are antigen independent?

6. How is it that each B cell produces a single antibody species (including allotypic exclusion) but the antibody response of the organism to a pathogen is quite diverse?

Topic Test 3: Answers

1. **False.** Pre-B cells must receive a signal via their pre-B BCR before they proceed to the next stage in development, but it cannot be antigen specific because foreign antigen is not present. Because surrogate light chains on all pre-B cell BCR are identical, the BCR do not have their ultimate antigen specificity.

2. **True.** Ig alpha and Ig beta are required for effective B-cell signaling.

3. **e.** B-cell differentiation begins with expression of *RAG-1*, *RAG-2*, and TdT, which allow somatic recombination to occur.

4. **a.** Pre-B cells proliferate before beginning L chain rearrangement.

5. Somatic recombination, negative selection, and receptor editing occur in the absence of foreign antigen.

6. The pathogen bears many epitopes recognized by many different B cells, each of which can differentiate into a clone of antibody-secreting plasma cells.

TOPIC 4: ISOTYPE SWITCHING

KEY POINTS

✓ *How is isotype switching both similar to and different from somatic recombination?*

✓ *What process allows B cells to produce IgM and IgD simultaneously?*

✓ *How do B cells produce either membrane or secreted IgM?*

Isotype switching increases the functional diversity of antibody molecules. It occurs after antigen stimulation and helper T-cell cytokine production. Each C gene segment except Cδ is preceded by an intron containing a **switch region sequence** different from the recombination signal sequences flanking V region segments; recombinases mediating isotype switching are not encoded by *RAG-1* and *RAG-2*. Rearranged VDJ_H is always expressed first with membrane Cμ in the developing B cell, with both membrane Cμ and Cδ in the mature B cell, and with secreted Cμ as the B cell begins responding to antigen. When the B cell receives the proper signals from antigen and cytokines to switch to IgG_3 production, for example, recombination occurs between switch regions Sμ and Sγ3, looping out intervening DNA including coding sequences for mu and delta chains. All isotype switching events are productive, because DNA splicing occurs within introns. Further switches to downstream isotypes may occur.

B cells produce IgM and IgD receptors simultaneously without isotype switching by **alternative mRNA splicing** (Figure 4.1). One mRNA encodes VDJ_H–Cμ, which is translated into membrane IgM; the other encodes VDJ_H–Cμ–Cδ, which is spliced to remove Cμ exons and translated into membrane IgD. Membrane and secreted forms of IgM are also produced by alternative splicing of shorter or longer mRNAs.

Topic Test 4: Isotype Switching

True/False

1. Isotype switching prevents the B cell from ever expressing all upstream isotypes.
2. Alternative mRNA splicing does not remove B-cell genetic material.

Multiple Choice

3. Isotype switching resembles somatic recombination because both processes
   a. Are catalyzed by the products of *RAG-1* and *RAG-2*.
   b. Are regulated by helper T-cell cytokines.
   c. Can result in stop codons in coding sequences.
   d. Occur in developing B cells in the bone marrow.
   e. Result in the irreversible loss of DNA from the B cell.
4. Alternative mRNA splicing
   a. Allows the B cell to improve its antigen-binding fit after antigen contact.
   b. Allows the B cell to make membrane IgM from the longer mRNA for secreted IgM.
   c. Can be used for the simultaneous production of any two Ig isotypes.
   d. Is a process by which a B cell can simultaneously synthesize mu and delta chains.
   e. Occurs in response to T-cell cytokines.

Short Answer

5. Why is isotype switching always productive?
6. What would be the fate of a B cell that switched isotypes to the pseudogene Cε?

Topic Test 4: Answers

1. **True.** DNA upstream (5′) of the switch site is looped out of the DNA.
2. **True.** Alternative mRNA splicing involves only message processing, not DNA.
3. **e.** Answers a, c, and d are true only for somatic recombination; b is true only for isotype switching.
4. **d.** Alternative mRNA splicing allows the B cell to simultaneously synthesize mu and delta chains; mRNA for other isotypes is not made simultaneously.
5. The switch sites are in introns, not exons, where they might cause frame shift mutations.
6. It would not be able to make either IgE antibody or the previous antibody isotype. Unless it received another signal allowing it to switch to a functional C_H gene, it would die without producing any more antibody.

CLINICAL CORRELATION: X-LINKED AGAMMAGLOBULINEMIA

X-linked agammaglobulinemia (XLA) is caused by the absence of Bruton's thymidine kinase (Btk) and results in nearly complete absence of serum antibodies and recurrent pyogenic and enteric infections. Children with XLA are usually diagnosed when maternal IgG levels fall after birth. In the absence of Btk, signaling via pre-B receptor to induce pre-B-cell maturation is very inefficient. Phenotypic analysis of circulating lymphocytes shows predominantly $CD3^+$ Ig^- cells; fewer than 0.1% are mu^+delta^+, compared with normal values of 5 to 15%. Bone marrow contains cells that are positive by immunofluorescence for membrane mu, VpreB, λ5, Ig alpha, and Ig beta but not membrane or cytoplasmic kappa, lambda, or delta chains. A different but related tyrosine kinase, Lck, is required for T-cell maturation. Children with XLA are given human gamma globulin to provide passive antibody protection.

DEMONSTRATION PROBLEM

A transgenic mouse was constructed by inserting recombined H and L chain genes encoding antibody to mouse H-Y antigen, expressed on membranes of all cells of male but not female mice. Describe B-cell development in male and female transgenic mice.

SOLUTION

The presence of rearranged H and L chain genes in the transgenic mice suppresses somatic recombination, so most B cells express transgenic BCR. In female mice, normal numbers of mature mu^+delta^+ B cells are present in marrow, circulation, and secondary lymphoid organs, all with identical H-Y antigen-specific BCR. In transgenic male mice there are normal numbers of pre-B cells, but B cells expressing transgenic IgM bind tightly to $H\text{-}Y^+$ cells in bone marrow and most die. A few immature B cells undergo receptor editing and emerge with BCR bearing V_L or V_H not specific for H-Y; these B cells mature and migrate to the secondary lymphoid organs, which are smaller than those in the female mice.

Chapter Test

True/False

1. Allelic exclusion could not be detected in inbred mice.
2. No antibody diversity is inherited from our parents.
3. Information controlling which gene segments can be recombined is present in the spacing between recombination signal sequences.
4. Introns between C_H domains are removed during somatic recombination.
5. Progression of B cells from pre-B to mature B requires expression of Btk.

Multiple Choice

6. Which is *not* a mechanism for generating diversity during B-cell development?
   a. Addition of extra bases during somatic recombination.
   b. Germline diversity in V, D, and J sequences.
   c. Imprecise joining of variable region gene segments.
   d. Polyclonal activation of B cells.
   e. Random combination of heavy and light chains into functional antibodies.
7. The insertion by TdT of one or more additional bases at a gene splice site is called
   a. Antigenic variation.
   b. Looping out.
   c. N region addition.
   d. Receptor editing.
   e. Sister chromatid exchange.
8. Somatic recombination of Ig gene segments
   a. Ensures that a mature B cell expresses Ig with a single isotype.
   b. Is carefully regulated so that self-specific B cells are never produced.
   c. Occurs in random order.
   d. Requires T cell help.
   e. Results in increased transcription of Ig genes.
9. What occurs first during B-cell development?
   a. Antigen binds BCR.
   b. H chain rearranges.
   c. IgM is expressed on the B-cell membrane.
   d. L chain rearranges.
   e. Promoter and enhancer sequences are brought closer together.
10. Which of the following is *not* true concerning the expression of L chain genes?
    a. Glycosylation occurs in the endoplasmic reticulum and Golgi region.
    b. Leader sequence is removed in the endoplasmic reticulum.
    c. Messenger RNA binds to ribosomes and is translated.
    d. Translation into protein occurs in the cytoplasm.
    e. V and J gene segments encode the variable region of light chain.

Short Answer

11. Why is it important for each B cell to express only one VDJ_H and one VJ_L?
12. What are the possible fates of the B cells generated in the bone marrow?
13. Name two enzymes uniquely associated with gene rearrangement in lymphocytes. What consequences would a deficiency in either of these genes have?
14. How could you count pre-B cells and mature B cells in the bone marrow?

Essay

15. Explain how somatic recombination results in generation of antigen-binding diversity for BCR and secreted antibodies.

Chapter Test: Answers

1. **T** 2. **F** 3. **T** 4. **F** 5. **T** 6. **d** 7. **c** 8. **e** 9. **b** 10. **a**

11. Clonal selection demands that each lymphocyte is specific for a single antigen, so that its stimulation results in a specific immune response.
12. B cells either die by apoptosis if they bind self-antigen in the marrow or they leave the marrow for the secondary lymphoid organs.
13. DNA recombinase, encoded by *RAG-1* and *RAG-2*, and TdT; severe combined immunodeficiency (SCID).
14. Stain the bone marrow with fluorescent anti-mu and anti-delta antisera. Cells with only mu chain are the pre-B cells; mu^+delta^+ cells are mature B cells.
15. Antigen-binding diversity begins with the many germline V, D, and J segments that encode V_H and V_L chains. Combinatorial diversity is generated by the random recombination of these segments into VJ_L and VDJ_H and by the ability of any V_H to pair with any V_L to bind antigen. Junctional diversity affecting predominantly CDR3 results from the imprecise joining of gene segments and from the addition of nucleotides at splice sites. Isotype switching allows these recombined variable sequences to be expressed as both membrane and secreted antibody and isotypes of different biological function.

Check Your Performance:

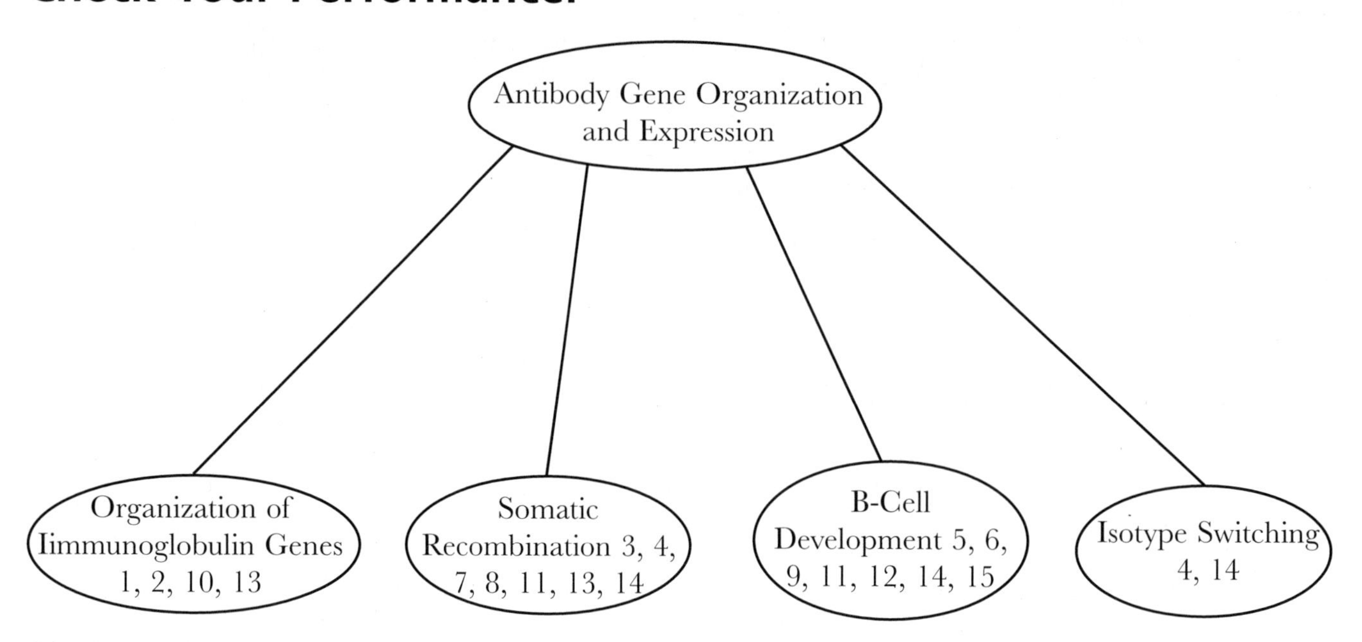

Check your understanding of this chapter by noting the number of questions for each topic you missed on the chapter test.

Major Histocompatibility Complex and Antigen Presentation

Antigens are presented to T cells on the cell-surface molecules that determine "self" or "non-self" for transplantation, called **human leukocyte antigens** (HLA) in humans and **H-2 antigens** in mice. They are encoded by genes of the major histocompatibility complex (MHC). MHC class I proteins are expressed on all nucleated cells in the body; cytotoxic T lymphocytes (T_C) bind endogenous peptide presented on infected cell class I and kill the infected cell. MHC class II proteins are expressed constitutively on B cells, dendritic cells, and thymic epithelial cells and can be induced on macrophages and human T cells. Helper T cells (T_H) bind exogenous peptide presented on antigen-presenting cell (APC) class II and produce cytokines to help B cells make antibodies and activate macrophages.

ESSENTIAL BACKGROUND

- **Eukaryotic cell structure**
- **Gene structure, function, and inheritance**
- **Endogenous and exogenous antigen (Chapter 1)**
- **Antibody structure (Chapter 3)**

TOPIC 1: MHC PROTEINS AND GENES

KEY POINTS

✓ *What is the structure of MHC class I and class II proteins?*

✓ *On what cells are MHC class I and class II expressed?*

✓ *How are MHC genes organized on the chromosome?*

✓ *How are MHC alleles inherited?*

✓ *How is MHC gene expression regulated?*

Figure 5.1 represents the structure of human HLA proteins. Class I MHC is a heterodimer of a membrane-bound alpha chain encoded by highly polymorphic genes in the MHC and noncovalently associated **beta$_2$-microglobulin** (**b$_2$M**) encoded by a nonpolymorphic gene on a different chromosome. Alpha chain and b$_2$M are members of the immunoglobulin (Ig) superfamily.

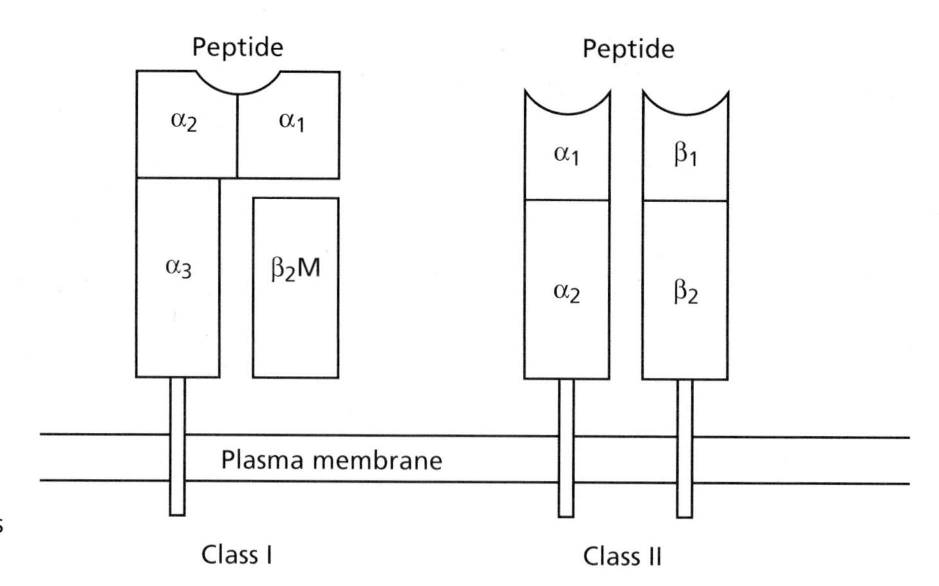

Figure 5.1. Structure of MHC class I and MHC class II proteins.

Antigen binds alpha1 and alpha2 domains; the amino acid sequence of these domains varies from allele to allele. Variability is maximized in the amino acids that contact antigen but is less than that for Ig V_H and V_L domains. Class I alpha chain folds so its variable region forms a cleft that holds an 8- to 10-amino acid peptide; the cleft is closed at the ends, limiting the size of the binding peptide. The alpha1 and alpha2 domains also bind T-cell receptor (TCR). CD8 binds class I alpha3 domain, which is invariant within a species. Three class I gene loci encode human HLA-A, B, and C or mouse D, K, and L alpha chains. MHC class II is a noncovalently bonded alpha–beta heterodimer called DP, DQ, and DR in humans and IA and IE in mice. Peptide antigen (13 to 18 residues) binds class II alpha1 and beta1 domains, which are variable; alpha1 and beta1 also bind TCR. Membrane-binding alpha2 and beta2 domains are invariant, and alpha2 binds CD4. Class II peptide-binding site resembles that of class I except that its ends are open so that longer peptides can bind.

Although numerous MHC alleles have been identified in humans, each individual has a limited number of MHC molecules with which to present many pathogen epitopes to T cells. Peptide binding to MHC is less specific than epitope binding to Ig or TCR; each MHC presents many different epitopes. Peptide must bind MHC with enough affinity to be retained on the plasma membrane and not exchange with soluble peptide. MHC molecules are unstable in the absence of peptide and are folded around peptide before transport to the plasma membrane. Peptide that binds class I lies extended along the antigen binding cleft, anchored at its amino and carboxyl termini to invariant sites at each end of the cleft. Anchor residues at two or three other positions along the peptide interact with residues in the MHC binding cleft. For a given MHC allele, all binding peptides have hydrophobic residues for one anchor site, for example, or all acidic amino acids. Anchor residues vary between MHC alleles, and residues not in the anchor positions can vary considerably, allowing many different peptides to be presented by a few class I alleles. TCR distinguishes different peptides by their conformation and by the class I conformation induced by their binding. Peptides that bind class II are at least 13 amino acids long and are not anchored by their amino and carboxyl termini. They stretch along the binding grove, with residues fitting into binding pockets along the edges of the cleft. Class II binding peptides also have a small number of anchor residues that must be of certain types to bind a particular allele; other residues in the peptide are less constrained.

Human MHC is a cluster of genes on a single chromosome. MHC is highly polygenic and polymorphic. Three human class I genes encode HLA-A, B, and C alpha chains. Class II genes for DP, DQ, and DR are arranged in pairs encoding alpha and beta chains. The class II region also encodes proteins for processing (LMP—low-molecular-weight proteins) and transporting (TAP—transporter associated with antigen processing proteins) cytoplasmic peptides into the endoplasmic reticulum (ER) for binding class I. MHC class III encodes complement proteins, cytokines tumor necrosis factor alpha and beta, enzymes required for steroid synthesis, heat shock proteins, and many unidentified proteins.

Numerous HLA alleles have been identified; because studies of some races have been limited, the numbers underestimate the polymorphism of the entire human species. For each HLA-A, B, and C, 95, 207, and 50 alleles have been identified. The alpha and beta chain alleles of HLA-DP, DQ, and DR number 12 and 80, 20 and 35, and 1 and 239, respectively. All alleles do not appear with equal frequencies. This is population diversity because each individual inherits a maximum of two alleles for each locus. MHC genes are expressed codominantly on cell membranes. The alpha chains of one class II allele can combine with beta chains of the same or the other allele for that locus, increasing the number of class II molecules that can be expressed. The set of MHC genes inherited from each parent is called a **haplotype**. Siblings have a one in four chance of inheriting identical haplotypes from both parents and are likely to provide the best match for transplantation. MHC polymorphism makes tissue matching between unrelated people very unlikely. Although MHC class I and class II are expressed constitutively in high amounts by a few cell types, especially dendritic and B cells, they are present in low amounts on most cells. Their expression, along with processing enzymes, transporter subunits, and chaperones to aid in their proper folding and association with peptide, is stimulated by interferon gamma (IFN gamma) produced early in viral infections.

Topic Test 1: MHC Proteins and Genes

True/False

1. MHC class I and class II are less antigen specific than Ig.
2. Each individual expresses all the diversity of MHC protein structure.

Multiple Choice

3. The MHC has
   a. Dozens of loci for class I and class II proteins.
   b. Genes that encode proteins associated with antigen processing.
   c. Only genes encoding class I and class II molecules.
   d. Regions encoding class III antigen-presenting molecules.
   e. Single loci for class I and class II proteins.
4. Both MHC class I and class II molecules are
   a. Composed of alpha and beta chains with variable and constant regions.
   b. Expressed constitutively on all nucleated cells.
   c. Expressed on the B-cell membrane.
   d. Part of the T-cell receptor for antigen.
   e. Synthesized in response to antigen processing.

Short Answer

5. Describe the antigen binding site of MHC class I.
6. What are the maximum number of class I and class II proteins expressed on a liver cell? On a macrophage?

Topic Test 1: Answers

1. **True.** There are fewer MHC molecules to present all the peptides to which T cells can respond; therefore, each MHC protein must be able to bind many different peptides.
2. **False.** Unlike Ig diversity, which is very high in each person, MHC diversity is much higher in the species than in the individual.
3. **b.** The human MHC has three loci each for class I and class II proteins. It also has genes for molecules required for processing and transporting peptides. Some class III molecules have immune functions, but they do not present antigen.
4. **c.** As APC, B cells express class II; all nucleated cells express class I.
5. Antigen binds the alpha1 and alpha2 class I domains. Alpha chain folds so that its variable region forms a cleft that holds an 8- to 10-amino acid peptide; ends of the cleft are closed.
6. The maximum number of class I proteins expressed on liver cells and macrophages is six: two alleles each of HLA-A, B, and C. Liver cells do not express class II. The number of class II proteins expressed on a heterozygous macrophage is at least 12: two alleles each of HLA-DP, DQ, and DR where the alpha and beta chains from the same allele are paired and the same number where the alpha and beta chains for each locus are paired with those of the other allele. For a person with DP alleles 1 and 2, for example, the DP proteins expressed would be $alpha_1beta_1$, $alpha_1beta_2$, $alpha_2beta_1$, and $alpha_2beta_2$. Because some class II loci have more than one beta chain, more alpha–beta combinations are possible.

TOPIC 2: ANTIGEN-PRESENTING CELLS

Key Points

✓ *What cells present antigen to T cells?*

✓ *Where are APC found?*

✓ *How does antigen presentation target the immune response?*

$CD4^+$ T_H recognize peptide antigen on membrane class II of macrophages, dendritic cells, and B cells. To activate T cells, APC must present antigen efficiently and express co-stimulatory molecules. Macrophages have receptors for bacterial surface carbohydrates and lipopolysaccharide and for complement and Fc to facilitate phagocytosis. Constitutive levels of class II and co-stimulatory molecules are low on macrophages, but bacterial antigens and inflammatory cytokines increase their expression.

Interdigitating dendritic cells are present in T-cell areas of secondary lymphoid tissues and thymus. They have high constitutive levels of class II and co-stimulatory molecules and extensive surface area. Dendritic cells are not phagocytic but are infected by many viruses. Infected dendritic cells present viral envelope proteins on class II to T_H and viral endogenous antigen on class I to T_C. Related cells with few class II and co-stimulatory molecules are present in skin and digestive epithelia and take up antigen by pinocytosis. During infection, these cells migrate to lymphoid tissues, increase their expression of class II and co-stimulatory molecules, and become dendritic cells.

Antigen cross-linking of B-cell receptor (BCR) stimulates endocytosis and antigen presentation by B cells. B cells have high levels of class II; co-stimulatory molecule expression increases after contact with bacterial carbohydrates, lipopolysaccharide, and T_H cytokines. B cells are present in secondary lymphoid organs where they are likely to contact T cells. B cells activate T cells at lower antigen concentrations than do macrophages.

T_H that see antigen on class II differentiate into two effector T_H phenotypes. Inflammatory (T_H1) cells secrete cytokines that stimulate macrophages to kill phagocytosed bacteria. Helper (T_H2) cells secrete cytokines that stimulate B cells to mature into antibody-secreting plasma cells and undergo isotype switching.

T_C recognize antigen on MHC class I. Because MHC class I is present on nearly all nucleated cells, virus-infected cells present virus peptides to T_C, which kill them. Because most nucleated cells do not express co-stimulatory signals for T-cell activation, T_H1 cytokines are required to fully activate T_C.

Topic Test 2: Antigen-Presenting Cells

True/False

1. Macrophages express high levels of class II and co-stimulatory molecules.
2. B cells present on class I only endogenous antigens that their BCR bind.

Multiple Choice

3. In addition to expressing membrane class II, an efficient APC must
   a. Be able to perform phagocytosis.
   b. Be present in the B cell areas of the lymphoid organs.
   c. Express co-stimulatory molecules.
   d. Have membrane antibody to bind antigen.
   e. Secrete IFN to kill the antigen.
4. T_H2 cells
   a. Are memory T cells.
   b. Secrete cytokines that activate B cells to produce antibody.
   c. Secrete cytokines that activate T_C to lyse infected cells.
   d. Secrete cytokines that activate macrophages to kill phagosomal pathogens.
   e. See antigen on class I.

Short Answer

5. How do macrophages and B cells differ in the antigens they present on class II?

6. How does antigen presentation target the immune response appropriately to exogenous and endogenous antigen elimination?

Topic Test 1: Answers

1. **False.** Class II and co-stimulatory molecule expression is low on unstimulated macrophages but is increased by bacterial antigens and inflammatory cytokines.
2. **False.** B cells take up exogenous antigen by endocytosis after BCR binding.
3. **c.** To become activated, T cells must bind specific peptide and receive a co-stimulatory signal.
4. **b.** T_H1 cells secrete cytokines that activate T_C to lyse infected cells and macrophages to kill endosomal pathogens. All T_H cells see antigen on class II.
5. Macrophages present exogenous peptides from any pathogen they engulf. B cells present exogenous peptides only from pathogens that bind BCR.
6. Because endogenous peptides are presented on target cell class I, they activate T_C to kill the targets. Exogenous antigen activates inflammatory and helper T_H cells, which stimulate the production of antibody and inflammation.

TOPIC 3: ANTIGEN PROCESSING AND PRESENTATION

KEY POINTS

✓ *In what cellular compartment of APC are exogenous antigens found?*

✓ *How are exogenous antigens processed and presented?*

✓ *In what cellular compartment of APC are endogenous antigens found?*

✓ *How are endogenous antigens processed and presented?*

Exogenous antigen is processed in the **endosomal processing pathway**. Bacteria, soluble protein antigens, antibody-coated viruses taken up by macrophages and B cells, virus envelope proteins from the plasma membrane of infected APC, and pathogens that live in phagosomes enter the endosomal pathway. Endosomes become increasingly acidic as they move into the cytoplasm, activating proteases that cut antigen into peptides. Class II alpha and beta chains are synthesized on the rough ER and transported into the lumen, where they assemble with the **invariant chain (Ii)**. Ii and class II form nine-chain trimeric complexes, with part of each Ii molecule occupying the class II peptide binding site. Ii allows class II to assemble in the absence of foreign peptide and blocks association with self and endogenous peptides present in the ER. Ii also directs class II transfer to an acidic compartment, **MIIC**, where Ii is degraded. The last part of Ii to dissociate from class II is **CLIP** (class II–associated invariant chain peptide), a short fragment occupying the peptide binding site. **HLA-DM**, a class II alpha–beta heterodimer not expressed on the cell surface, facilitates CLIP removal and peptide binding in MIIC. Class II which does not bind peptide when CLIP dissociates is rapidly degraded; in the absence of infection, APC class II bind self-peptides, including class II peptides. Class II–peptide complex is transported to the plasma membrane. The complex is very stable, ensuring that the APC presents its own exogenous peptides. Class II proteins shuttle between the membrane and endosomal compartments where they bind new exogenous peptide or are degraded.

A virus-infected cell synthesizes virus proteins on cytoplasmic ribosomes. In the **cytosolic processing pathway**, cytoplasmic proteins are degraded to peptides in **proteasomes**, cylindrical arrays of proteolytic enzymes with their active sites toward the center of the cylinder. Two proteases encoded in MHC class II (LMP2 and LMP7) and a third subunit not encoded in MHC are synthesized in response to IFN gamma. These inducible proteases replace constitutive proteasomal enzymes and produce peptides with basic and hydrophobic carboxyl terminal residues preferred as anchor residues in class I peptide binding sites and for transport to the ER. **TAP-1** and **TAP-2** are ER membrane proteins with ATP-binding domains on the cytosolic side and hydrophobic transmembrane domains. The TAP-1–TAP-2 complex transports cytosolic peptides into the ER lumen with expenditure of ATP. Both TAP molecules are required for membrane expression of class I.

Newly synthesized, partly folded, class I alpha chain binds the chaperone **calnexin** in the ER lumen. When b_2M binds class I, calnexin dissociates and class I alpha–b_2M forms a complex with **calreticulin**, **tapasin**, and TAP. Peptide binding releases class I from TAP, and the class I–peptide complex is transported to the plasma membrane; unbound peptides return to the cytoplasm. In uninfected cells, membrane class I bears endogenous self-peptides. Viruses that interfere with TAP function or class I transport to the membrane evade destruction by T_C.

Topic Test 3: Antigen Processing and Presentation

True/False

1. Invariant chain is the constant part of the MHC class II peptide binding site.
2. In the absence of infection, class I without bound peptide is expressed.

Multiple Choice

3. Exogenous antigen is processed
   a. After presentation by APC.
   b. By nearly every nucleated cell.
   c. By the cytosolic processing pathway.
   d. In the presence of b_2M.
   e. In acidified endosomes.
4. After virus infection, virus peptides processed in proteasomes are likely to be presented because
   a. Class I is synthesized in response to virus infection.
   b. Proteosomal enzymes that produce shorter peptides are synthesized in response to virus infection.
   c. TAP-1 and TAP-2 specifically bind virus peptides.
   d. Virus peptides bind better to class I than peptides from self-proteins.
   e. Virus infection induces expression of proteases that cut proteins at sites favorable to TAP-1/TAP-2 and class I binding.

Short Answer

5. How can a cell present self-MHC peptides on MHC class II?
6. Which pathogens require the cytosolic processing pathway for detection?

Topic Test 3: Answers

1. **False.** Invariant chain is a separate molecule that binds the class II peptide binding site, facilitates folding, and blocks binding of endogenous and self-peptides.
2. **False.** In the absence of virus infection, class I binds self-peptide.
3. **e.** Exogenous antigen is processed in acidified endosomes. Answers b, c, and d are true for endogenous antigen.
4. **e.** IFN-induced proteases produce peptides with carboxyl terminal residues that bind more efficiently to TAP-1/TAP-2 and class I.
5. MHC molecules are degraded rapidly in the absence of peptides. Intact class II in an APC without foreign peptide in MIIC binds degraded class II peptides.
6. The cytosolic processing pathway is particularly important for recognition of nonenveloped viruses, which express no virus proteins on the infected cell membrane.

TOPIC 4: MHC AND IMMUNE RESPONSIVENESS

KEY POINTS

✓ *What evidence led to the discovery of MHC restriction?*

✓ *Why are T-cell MHC restricted in their ability to see antigen?*

✓ *What are minor histocompatibility antigens?*

When inbred mice were used to investigate immune activation, it was discovered that the ability to recognize antigens (particularly haptens) was genetically linked to MHC. Some strains of inbred mice were nonresponsive to particular haptens; no IgG antibody was produced to that hapten in that strain, although in other strains of mice the same hapten–carrier complex was immunogenic. When **syngeneic** (matched at all MHC class I and class II loci) mouse T cells, B cells, and macrophages were mixed with antigen in vitro, antibody to the antigen was produced; when **allogeneic** cells (differing at MHC loci) were used, the immune response was absent. T cells are **MHC restricted** in their ability to see antigen; they only recognize antigen presented by syngeneic MHC on APC.

T_H ($CD4^+8^-$) cells bind exogenous antigen on class II of APC: dendritic cells, macrophages, and B cells. TCR binds peptide and class II alpha1 and beta1 domains; CD4 binds alpha2. T_C ($CD4^-8^+$) cells recognize antigen on class I of cells infected with endogenous pathogens. TCR binds peptide plus class I alpha1 and alpha2 domains; CD8 binds alpha3. Exogenous and endogenous antigens are processed by different pathways that transport them to the appropriate MHC proteins for presentation on the membrane. Developing T cells are educated (selected) in the thymus to bind self-MHC. A few people have been discovered with **bare lymphocyte syndrome**, a partial or complete deficiency in class I or class II proteins. People with bare lymphocyte syndrome are more susceptible to viral and opportunistic infections. Symptoms range from none to severe combined immune deficiency, depending on the number of MHC loci that can be expressed.

A genetic linkage exists between HLA type and several diseases. Insulin-dependent diabetes occurs four times as often in people with HLA-DR4 as in people without the DR4 allele; people with HLA-A3 have a sevenfold increased risk of developing idiopathic hemochromatosis (excessive iron levels); and ankylosing spondylitis (arthritis of spinal vertebrae) has a B27-linked risk factor of 87. Proteins encoded by genes closely linked with MHC may be responsible for these diseases, or disease may be caused by viruses that use particular MHC alleles to enter host cells. Molecular mimicry between pathogen and MHC epitopes may exist, so that antibodies made against the pathogen bind self-MHC and cause autoimmunity; or thymic education in people with certain alleles may fail to delete self-specific T cells.

In addition to MHC class I and class II, **minor histocompatibility antigens** induce a weaker graft rejection response. Some are tissue specific (skin) or sex specific (male). Some are proteins encoded by viral DNA integrated with host cell DNA (Mls antigens of mice). Others are foreign peptides bound to class I and class II that occasionally trigger graft rejection episodes between tissues of identical twins.

Topic Test 4: MHC and Immune Responsiveness

True/False

1. $CD4^+$ T cells see antigen on self-MHC class II but not on self-MHC class I.
2. Someone with bare lymphocyte syndrome who expressed no MHC proteins would die in infancy.

Multiple Choice

3. T cells are MHC-restricted in their ability to respond to antigen because
   a. All antigen must be processed and presented to activate lymphocytes.
   b. During an infection, all cells in the body present antigen on MHC class I.
   c. MHC binds antigen more specifically than TCR does.
   d. TCR must recognize both antigen and MHC molecules.
   e. They use MHC as their TCR.
4. Linkage of a disease to an HLA allele means that
   a. Everyone with that allele will eventually get the disease.
   b. People with that allele have a higher risk for the disease.
   c. The MHC protein encoded by that allele is defective.
   d. The allele will eventually disappear from the population.
   e. None of the above

Short Answer

5. Compare the chances of an inbred mouse being a nonresponder to a hapten with the chances of a person being a nonresponder to a pathogen.
6. What are minor histocompatibility antigens?

Topic Test 4: Answers

1. **True.** $CD4^+$ T cells are restricted to self-MHC class II.
2. **True.** Total lack of MHC would prevent T-cell activation and development, and without T cells there would be no help for antibody production or inflammatory cytokines to activate macrophages. This individual would die without extreme medical intervention.
3. **d.** T cells are educated to recognize self-MHC, so their response is restricted to peptide on self-MHC. Antigen presentation is not required for B-lymphocyte activation.
4. **b.** Most people with a given MHC allele do not get the disease that is linked to it, but their risk is higher. Most of these diseases do not interfere with reproductive ability (population selection). Selective pressure favors the retention of alleles useful for presenting common pathogen peptides.
5. The mouse's chances of being a nonresponder are much higher, because it is homozygous for MHC and because the hapten has only one epitope that might not be presented by any of the alleles. A human is heterozygous, so he or she has more MHC alleles with which to bind antigen, and pathogens have many proteins that can be processed to numerous peptides for presentation.
6. Minor histocompatibility antigens induce a weaker rejection response than MHC. They are tissue- or sex-specific self-antigens, endogenous viral proteins, or foreign peptides bound to class I and class II.

CLINICAL CORRELATION: TAP-1 DEFICIENCY

TAP-1 deficiency results in elevated susceptibility to virus infections usually apparent in early childhood. Circulating T lymphocytes are 95% $CD4^+$, compared with a normal 67%. Class I on leukocytes is reduced about 90%. Without TAP-1, virus peptides are not transported to the ER to bind class I; without membrane class I, $CD8^+$ T_C development does not occur. Serum IgG is produced to childhood vaccinations; B-cell and neutrophil counts and complement functions are normal. Class II is expressed and $CD4^+$ T cells develop properly. T_H1 inflammatory cells and T_H2 helper cells provide normal immunity against exogenous antigens.

DEMONSTRATION PROBLEM

To understand MHC gene expression, you mate two inbred mice. The first, from strain A, has the MHC class I genotype H-2 D^a K^a and the class II genotype IA alpha^{a}beta^a IE alpha^{a}beta^a; it has a defective gene for b_2M. The second, from strain B, has the genotype H-2 D^b K^b IA alpha^{b}beta^b IE alpha^{b}beta^b and a normal b_2M gene; the gene for IE beta^b contains a premature stop codon. Describe the H-2 phenotype of B cells from each parent and from the F_1 progeny.

SOLUTION

B cells are APC and constitutively express all alleles of both class I and class II proteins. Each parent is homozygous; the progeny will be heterozygous. The A strain parent cannot express any class I because it cannot make b_2M. It expresses IA $alpha^a beta^a$ and IE $alpha^a beta^a$. The B strain parent has a functional b_2M and expresses H-2 D^b and K^b. It also expresses IA $alpha^b beta^b$ but not IE because the IE beta chain is defective. The F_1 progeny express H-2 D^a, K^a, D^b, and K^b because b_2M encoded by the B strain parent can be used to assemble both alleles of D and K. Expressed class II proteins are IA $alpha^a beta^a$, $alpha^a beta^b$, $alpha^b beta^a$, and $alpha^b beta^b$ and IE $alpha^a beta^a$ and $alpha^b beta^a$; IE $beta^b$ is not expressed.

Chapter Test

True/False

1. If a family has five children, two of them will always have the same MHC genotype.
2. Peptide binding to TCR is influenced by both its own conformation and the conformation of the MHC protein to which it is bound.
3. Calnexin, calreticulin, and tapasin protect the peptide binding site of class I from CLIP binding.
4. Virus peptides are never presented on B-cell class II.
5. HLA-DM is a human class II minor histocompatibility antigen.

Multiple Choice

6. All of the following are associated with the expression of class I molecules *except*
   a. Antigen peptide presentation on membrane MHC class I to T_C.
   b. Graft rejection.
   c. Increased risk of certain autoimmune diseases.
   d. Lysis of virus-infected cells.
   e. Stimulation of antibody production.
7. Humans inherit from each of their parents
   a. A random set of MHC class I, class II, and class III genes.
   b. Enough diversity in MHC to present epitopes from most pathogens.
   c. Enough diversity in MHC to present every possible antigen epitope.
   d. Genes for alpha and beta chains that can be recombined to increase their diversity.
   e. The same MHC class I and class II genes as their siblings.
8. The alpha chain of HLA-DR
   a. Can be expressed with the beta chain of any MHC molecule.
   b. Can be expressed with the beta chain of any MHC class II molecule.
   c. Can be expressed with the beta chain of any MHC class II DR molecule.
   d. Must be expressed with b_2M.
   e. Must be expressed with the beta chain of class II DR from the same chromosome.
9. Signaling to a T_C that a liver cell is infected with hepatitis virus depends on
   a. Binding of Ii to class I until peptide is loaded.
   b. Binding of TCR on the cytotoxic T cell to class II on the infected cell.
   c. Binding of the processed antigen to liver cell class I.

d. Processing the hepatitis virus peptides to the correct size and anchor residues in the endosomal pathway.
e. Both c and d

10. To have pathogen peptide plus class II molecules expressed on host cell membranes, all of the following are required *except*
a. b_2M.
b. CLIP.
c. HLA-DM.
d. HLA-DR, DP, and DQ alpha chains.
e. Ii.

Short Answer

11. Why might certain MHC alleles be associated with certain diseases?
12. Why can successful blood transfusions occur between people who are matched only for ABO and Rh antigens and not MHC?
13. What is the fewest monoclonal antibodies you would need to detect all human class I MHC molecules?
14. What factors regulate the expression of MHC on the cell membrane?

Essay

15. Describe antigen processing and presentation of antigens from Epstein-Barr virus (EBV), which has infected B cells.

Chapter Test: Answers

1. **F** 2. **T** 3. **F** 4. **F** 5. **F** 6. **e** 7. **b** 8. **c** 9. **c** 10. **a**

11. Proteins encoded by genes closely linked with MHC may be responsible, or disease may be caused by viruses that use particular MHC alleles as receptors. Antigenic mimicry may exist between pathogen and MHC allele, or thymic education in people with these alleles may fail to delete self-reactive T cells.
12. Erythrocytes are not nucleated and do not express MHC proteins.
13. One; to detect all human class I MHC molecules, use anti-b_2M.
14. MHC expression depends on the presence of functional genes for class I, class II, and b_2M. It also depends on functional genes for processing, transporter, and assembly molecules and molecules that help APC internalize antigen: receptors for bacterial cell-surface molecules, Fc receptor, and complement receptors. Expression is controlled by regulatory molecules in particular cell types, cytokine signals, and molecules made by some pathogens that block expression of their peptides on infected cells.
15. EBV is an enveloped herpes virus that infects B cells using complement receptor CR2 for entry. Using the cytosolic processing pathway, infected B cells present EBV proteins on class I to T_C cells, which lyse the infected B cells. Virus particles and proteins released

from the lysed B cells bind to EBV-specific BCR and are internalized by endocytosis, processed, and presented on class II to T_H. The T_H secrete cytokines to stimulate B-cell proliferation, differentiation, and antibody secretion. Infected B cells also present envelope proteins using the endosomal processing pathway and class II.

Check Your Performance:

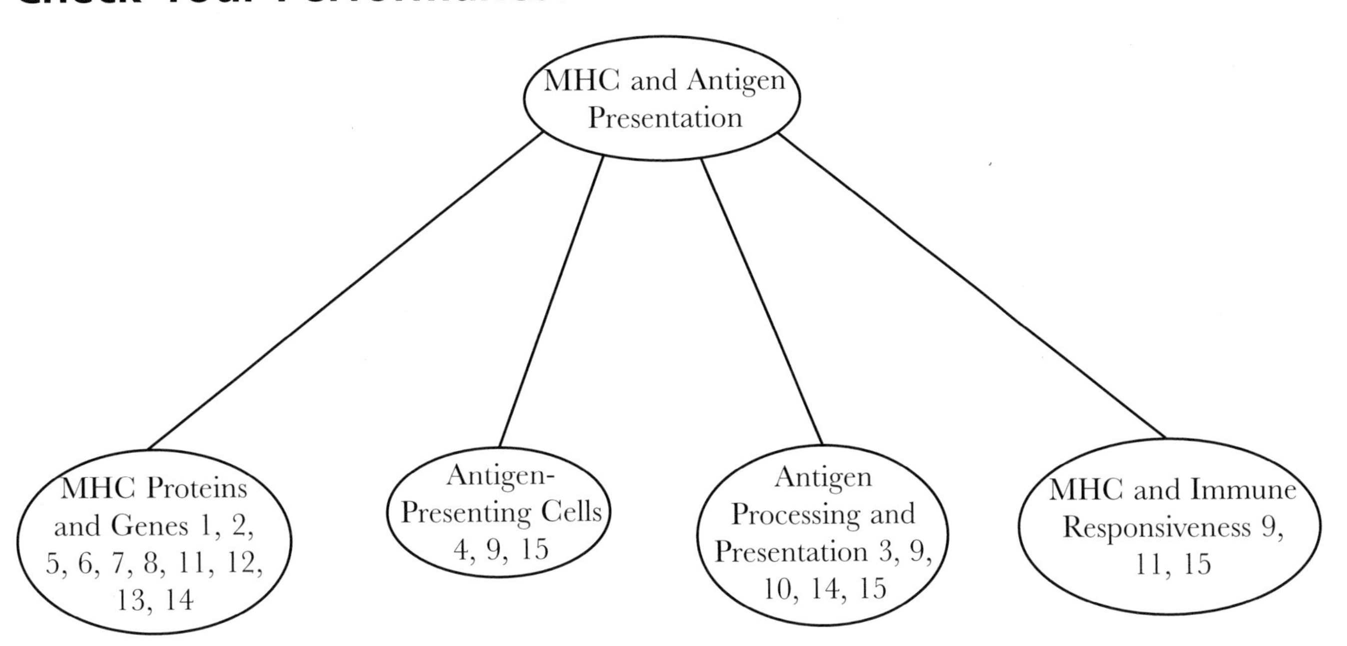

Check your understanding of this chapter by noting the number of questions for each topic you missed on the chapter test.

T-Cell Antigen Recognition

The absence of a secreted form of the T-cell receptor (TCR) and the requirement for recognition of both peptide and major histocompatibility complex (MHC) made isolation and characterization difficult. As T cells develop in the thymus, TCR gene segments are rearranged to produce unique TCR. T cells are then screened for their ability to bind self-peptide on self-MHC, and only those that bind with the appropriate affinity leave the thymus for the periphery.

ESSENTIAL BACKGROUND

- **Immunoglobulin (Ig) structure (Chapter 3)**
- **Ig gene organization and expression (Chapter 4)**
- **MHC structure and function (Chapter 5)**

TOPIC 1: TCR STRUCTURE AND ANTIGEN RECOGNITION

KEY POINTS

✓ *What is the structure of the T-cell antigen receptor molecule (TCR)?*

✓ *How does TCR structure correlate with antigen recognition?*

✓ *What co-receptors must be present with TCR on the T-cell membrane?*

Antigen-specific receptors on T cells are similar to B-cell receptor (BCR) (**Figure 6.1**). TCR belongs to the Ig superfamily. Each chain has a variable and a constant region; complementarity determining regions (CDR) define antigen-binding specificity and framework regions hold CDR in place. TCR is encoded in gene segments that undergo somatic recombination during T-cell development to generate antigen-binding diversity. Each T cell bears a single specificity and configuration of TCR. TCR is not secreted.

TCR is a heterodimer composed of alpha and beta chains or, on a minority of T cells, gamma and delta chains. The two chains are disulfide bonded just outside the plasma membrane in a short extended stretch of amino acids resembling Ig hinge; TCR have very short cytoplasmic tails. Both alpha and beta chains are glycosylated. Each TCR has a single binding site for antigen containing CDR3. CDR1 and CDR2 bind MHC. Antigen-binding affinity is lower than that of Ig for native antigen, but binding of co-receptor CD4 or CD8 to MHC increases the T-cell binding avidity. CD4 and CD8 also signal the T cell to become activated.

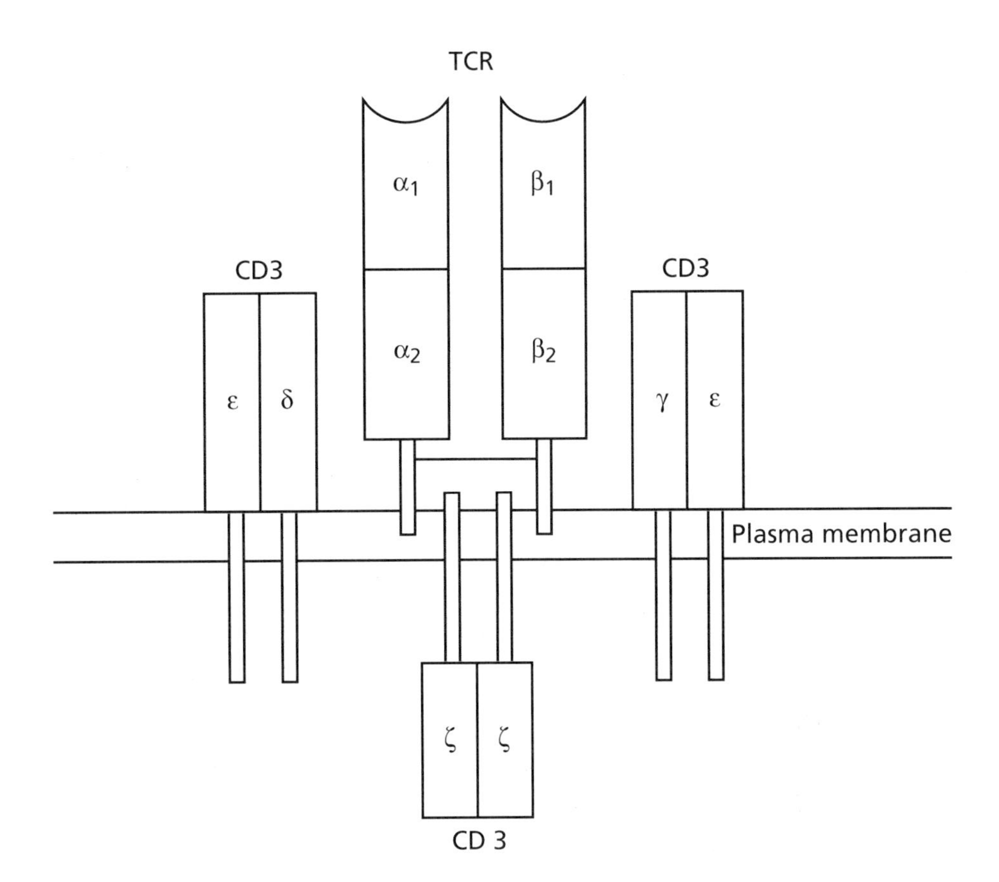

Figure 6.1 T-cell receptor structure.

TCR is expressed on the T-cell membrane with a signal transduction complex, CD3, also called the invariant TCR chains because CD3 molecules on all T cells are formed from identical subunits. CD3 is composed of three dimers: gamma epsilon, delta epsilon, and either two zetas or a zeta eta heterodimer. CD3 gamma and delta chains are not the same molecules found in the gamma delta TCR. The gamma, delta, and epsilon chains have negatively charged transmembrane regions that form salt bridges with positively charged transmembrane regions of TCR alpha and beta chains. Most of the zeta chain is cytoplasmic and transmembrane. Like Ig alpha/Ig beta, CD3 molecules have cytoplasmic ITAMs (immunoreceptor tyrosine activation motifs) that associate with tyrosine kinases after antigen binding to initiate a cytoplasmic cascade of activation signals.

CD4 is a monomeric protein with four Ig-like domains; the two most distal membrane domains bind the class II beta2 domain. CD8 is a disulfide-linked dimer; its alpha and beta chains each have one Ig-like domain with a long extended region connecting it to the transmembrane region. CD8 binds the alpha3 region of MHC class I. The cytoplasmic tails of both CD4 and CD8 associate with Lck to initiate signal transduction.

Topic Test 1: TCR Structure and Antigen Recognition

True/False

1. TCR most closely resembles the Fab region of Ig.
2. CD4 and CD8 are signal transducing molecules for T-cell activation.

Multiple Choice

3. Which of the following is *not* true about TCR?
   a. All TCRs on a particular T cell have identical idiotypes.
   b. CDR3 of TCR has the most sequence variability from molecule to molecule.
   c. TCR has binding sites for both antigen and self-MHC.
   d. TCR is a disulfide-bonded heterodimer.
   e. The alpha beta or gamma delta isotype of TCR determines the biological function of its secreted form.

4. Which of the following properties are *not* shared by TCR and BCR?
   a. Avidity is increased by the presence of two antigen-binding regions.
   b. Antigen-binding diversity is generated through gene rearrangement.
   c. Folding of protein domains is maintained by intrachain disulfide bonds.
   d. Membrane expression and lymphocyte activation by antigen require receptors to be associated with signal transduction molecules.
   e. Receptor antigen-binding sites are formed from two polypeptide chains.

Short Answer

5. What part of the TCR is responsible for its MHC-binding specificity?

6. Could TCR be purified on an affinity column of antigen peptide?

Topic Test 1: Answers

1. **True.** The chains of the T-cell heterodimer each have one variable and one constant region, with the two variable regions forming the binding site and the two constant regions disulfide bonded to each other. The primary difference between TCR and Fab is the transmembrane regions at the carboxyl termini of the TCR.

2. **True.** CD3, CD4, and CD8 transduce antigen/MHC binding signals into the T-cell cytoplasm.

3. **e.** TCR is not secreted from the T cell.

4. **a.** TCR have a single binding site for antigen; BCR are bivalent.

5. The MHC-binding specificity of TCR is thought to be primarily in the CDR1 and CDR2 regions of the molecule.

6. TCR could not be purified on an affinity column of antigen peptide, because its binding specificity is for a peptide–MHC complex and binding affinity to peptide alone would not be high enough.

TOPIC 2: PRODUCING A FUNCTIONAL TCR MOLECULE

KEY POINTS

✓ *How are TCR genes arranged in the DNA?*

✓ *How is generation of antigen-binding diversity different for TCR and for BCR?*

✓ *How much diversity can be generated in alpha beta TCR molecules?*

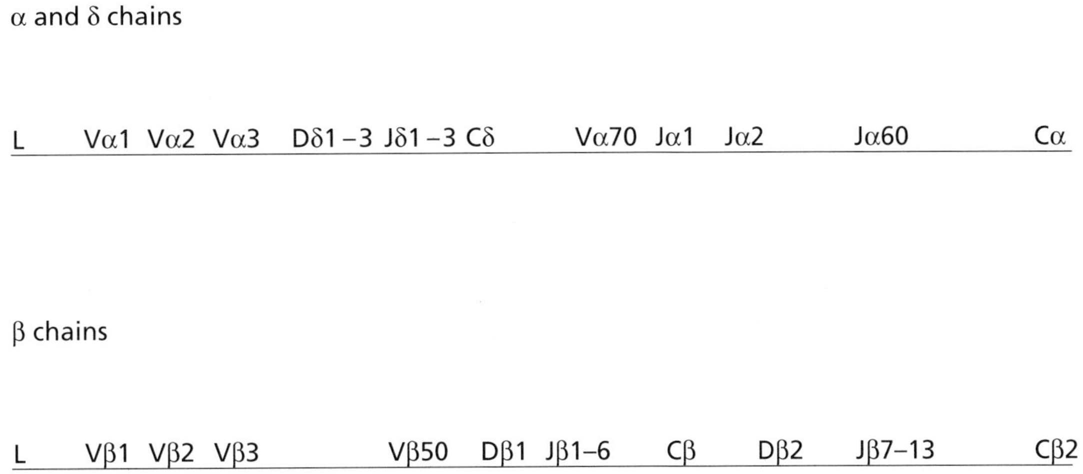

α and δ chains

L Vα1 Vα2 Vα3 Dδ1–3 Jδ1–3 Cδ Vα70 Jα1 Jα2 Jα60 Cα

β chains

L Vβ1 Vβ2 Vβ3 Vβ50 Dβ1 Jβ1–6 Cβ Dβ2 Jβ7–13 Cβ2

Figure 6.2 Arrangement of TCR genes. Vδ gene segments are intermingled with Vα segments.

TCR gene organization is similar to that for Ig genes (**Figure 6.2**), with TCR genes located in clusters on two (human) or three (mouse) different chromosomes. Vα and Vγ regions of TCR are encoded by V and J segments. Humans have approximately 70 different Vα, 60 different Jα, and a single Cα segment. A cluster of 12 Vγ is followed by 3 Jγ with Cγ1 and 2 Jγ with Cγ2. Vβ and Vδ chain regions are encoded by V, D, and J gene segments. Humans have approximately 50 Vβ and two clusters containing 1 Dβ, 6 to 7 Jβ, and 1 Cβ segment. Three Vδ, Dδ, and Jδ segments and a single Cδ segment have been counted; the Vδ segments are interspersed with Vα segments. The C region has separate gene segments for constant, hinge, transmembrane, and cytoplasmic domains.

Gene segments for TCR are flanked by the same recombination signal sequences as Ig gene segments, and RAG-1 and RAG-2 recombinases and terminal deoxynucleotidyl transferase (TdT) are required for somatic recombination. Joining regions for Vα and Jα and for Vβ, Dβ, and Jβ occur in CDR3, whereas CDR1 and CDR2 sequences are encoded within Vα and Vβ. Nucleotides are added to the junctions between Vβ, Dβ, and Jβ and between Vα and Jα. Generation of antigen-binding diversity for TCR, therefore, depends on the same combinatorial and junctional mechanisms used for Ig diversity. Somatic hypermutation does not seem to be an important diversity mechanism for TCR.

Topic Test 2: Producing a Functional TCR Molecule

True/False

1. The arrangement of alpha chain gene segments most closely resembles that of kappa chain.
2. Cα and Cβ are each encoded in a single exon.

Multiple Choice

3. Rearrangement of both TCR and BCR gene segments does *not*
   a. Generate diversity by recombination of many germline V, D, and J segments.
   b. Lead to CDR3 being the most hypervariable region.

c. Require RAG-1, RAG-2, and TdT expression.
d. Result in allelic exclusion of membrane receptors.
e. Result in class switching after antigen stimulation of mature lymphocytes.

4. T cells use all of the following for generating TCR diversity *except*
   a. Addition of nucleotides not encoded in the DNA.
   b. Combinatorial association of chains.
   c. Combinatorial association of exons.
   d. Large germline pool of gene sequences.
   e. Somatic hypermutation.

Short Answer

5. Assuming all gene segments could be used with equal frequency and ignoring the effects of imprecise joining and nucleotide additions, calculate how many possible Vα could be produced from the human alpha chain gene segments. Perform the same calculation for Vβ.
6. Compare the potential diversity in alpha beta TCR with that in BCR.

Topic Test 2: Answers

1. **True.** The alpha chain genes are arranged in a group of Vα, followed by Jα, and a single Cα. However, the gene segments for delta chain are within the Vα segment region.
2. **False.** Cα and Cβ are each encoded in four exons: constant, hinge, transmembrane, and cytoplasmic.
3. **e.** Class switching does not occur after antigen stimulation of the mature T lymphocytes.
4. **e.** T-cell receptors do not undergo somatic hypermutation.
5. Combinatorial variability for alpha chain is $70 \times 60 = 4{,}200$ possible Vα. The same calculation for Vβ is $50 \times 2 \times 13 = 1{,}300$ possible Vβ.
6. If any alpha chain can be used with any beta chain, the theoretical combinatorial diversity for alpha beta TCR would be $4{,}200 \times 1{,}300 = 5.46 \times 10^6$. This compares with about 2.9×10^6 possible Ig $V_H V_L$ regions.

TOPIC 3: THYMUS EDUCATION OF T LYMPHOCYTES

Key Points

✓ *At what point in differentiation do cells become committed to becoming T cells?*

✓ *How are T cells positively and negatively selected in the thymus?*

✓ *What is the importance of thymic education for T cells?*

Lymphoid progenitors that have developed from hematopoietic stem cells in the bone marrow migrate to the thymus to complete their antigen-independent maturation into functional T cells. In the thymus, T cells develop their specific T-cell markers, including TCR, CD3, CD4 or CD8, and CD2, and undergo thymic education through positive and negative selection. The thymus is a multilobed organ composed of cortical and medullary areas surrounded by a capsule. T-cell precursors enter the subcapsular cortical areas, where they proliferate. When they begin to express CD2 but have not yet rearranged their TCR genes, they are $CD2^+$ $CD3^-$ double-negative ($CD4^-$ $CD8^-$) cells. Of the double-negative cells in the thymus, about 20% rearrange gamma delta TCR, about 20% develop homogenous alpha beta TCR, and about 60% commit to becoming the majority of mature alpha beta T cells. These cells then express the adhesion molecule CD44 and the alpha chain of interleukin-2 receptor (CD25). $CD44^{low}$ $CD25^+$ double-negative T cells rearrange TCR beta chain, beginning with Dβ–Jβ joining and then Vβ–DJβ joining. The chances of successful beta chain rearrangement are increased by the presence of two DJCβ gene clusters. Productive rearrangement of beta chain results in its expression on the T cell membrane with CD3 and a **surrogate alpha chain**, **pre-Tα**. Membrane CD25 is lost; the cells stop rearranging beta chain, undergo a period of proliferation, and express both CD4 and CD8, becoming double-positive ($CD4^+CD8^+$) T cells. Double-positive cells reexpress RAG-1 and RAG-2 to rearrange their alpha chain genes. Alpha chain rearrangement can occur on both chromosomes and continues until the cell undergoes selection or dies; T cells are not allelically excluded for alpha chain. However, even cells with two different TCR have only one that can bind self-MHC with enough affinity to pass positive selection.

Double-positive alpha beta TCR^{low} cells must successfully undergo positive and negative selection before leaving the thymus. **Positive selection** occurs when double-positive T cells bind cortical epithelial cells expressing class I or class II. Positive selection determines whether the T cell will become a helper T cell (T_H) or a cytotoxic T lymphocyte (T_C). Selection with class I produces a $CD8^+$ T cell, whereas selection with class II yields a $CD4^+$ T cell. If the cell fails to bind, it can undergo further alpha chain rearrangements. Most T cells fail positive selection and die after 2 to 3 days in the thymic cortex. If TCR binds MHC, the T cell survives positive selection and migrates further into the corticomedullary junction where it encounters bone marrow-derived macrophages and dendritic cells with high levels of MHC self-peptide. T cells that bind self-peptide–MHC with high affinity at this stage undergo **negative selection** and die by apoptosis. Cells that are not negatively selected leave the thymus as single positive T_H ($CD2^+3^+4^+8^-$ alpha beta$^+$) or T_C ($CD2^+3^+4^-8^+$ alpha beta$^+$).

Gamma delta and alpha beta T cells arise from common precursors during TCR development. Beta, gamma, and delta TCR genes rearrange nearly simultaneously in the double-negative T cell; if the cell rearranges gamma and delta before it rearranges beta, it becomes a gamma delta T cell. If beta chain successfully rearranges first, the cell becomes an alpha beta T cell, because rearrangement of alpha chain usually deletes delta gene segments. Gamma delta T cells do not express CD4 and many of them do not express CD8, so gamma delta T cells do not undergo education for MHC restriction like alpha beta T cells. Gamma delta T cells develop in nude mice, which have no thymus. Because alpha chain genes do not rearrange until a few days after the other chains, gamma delta T cells appear first during mouse fetal development. The earliest gamma delta receptors expressed in mice are very homogenous. Gamma delta T cells initially home to skin, later to reproductive epithelium, and finally with more diverse gamma delta TCR to all secondary lymphoid tissue; their functions are unclear.

Topic Test 3: Thymus Education of T Lymphocytes

True/False

1. A T cell that fails positive selection dies immediately in the thymus.

2. A T cell that cannot productively rearrange alpha and beta chain genes will go on to rearrange gamma and delta chain genes.

Multiple Choice

3. Which of the following is *not* associated with negative selection?
   a. Apoptosis of thymocytes
   b. Bone-marrow derived antigen-presenting cells (APC).
   c. Recognition of self-MHC molecules.
   d. Recognition of foreign antigen.
   e. Self-tolerance.

4. Positive thymic selection
   a. Involves a mature single positive $CD4^+$ or $CD8^+$ T cell.
   b. Is successful for most thymocytes.
   c. Is required for self-MHC restriction of T cells.
   d. Results in self-tolerance of T cells.
   e. Occurs mostly in the medulla of the thymus.

Short Answer

5. How are early mouse gamma delta T cells different from later gamma delta T cells?

6. Because T cells are negatively selected, how can autoimmunity exist?

Topic Test 3: Answers

1. **False.** A T cell that fails positive selection can live for several days and continue to rearrange different alpha chain gene segments that might bind self-MHC.

2. **False.** A T cell becomes a gamma delta T cell only if it rearranges both gamma and delta before alpha and beta are rearranged. Rearrangement of alpha deletes delta chain genes.

3. **d.** Recognition of foreign antigen is not associated with negative selection, which occurs in the thymus before antigen contact. MHC in the thymus bears self-peptides.

4. **c.** Positive thymic selection is responsible for self-MHC restriction; only T cells that recognize self-class I or class II get the signal to survive.

5. Early mouse gamma delta T cells are very restricted in their $V\gamma$ and $V\delta$ segments and they home to specific regions of the body. Later, gamma delta T cells are more diverse in their TCR antigen specificity and are found in all secondary lymphoid organs.

6. T cells are negatively selected to self-peptides that appear in the thymus. This does not prevent T cells from recognizing peptides from proteins found only in certain organs or cells or foreign peptides that may cross-react enough with self to activate those T cells.

TOPIC 4: ALLOREACTIVITY OF T LYMPHOCYTES

KEY POINTS

- ✓ *How do T cells recognize foreign MHC molecules?*
- ✓ *Why is the frequency of alloreactive T cells so high for a particular MHC allele?*
- ✓ *What is the clinical significance of alloreactivity of T cells?*

T cells are positively and negatively selected in the thymus to be restricted to self-MHC. However, up to 5% of T cells react against cells bearing allogeneic MHC, compared with a much lower frequency of T cells that react against a particular foreign peptide–self-MHC complex. Alloreactivity of T cells is responsible for graft rejection between MHC-mismatched individuals. $CD8^+$ T cells respond to foreign class I and $CD4^+$ T cells respond to foreign class II. Some T cells may be specific for peptides on foreign MHC and may bind tightly enough to become activated even though the MHC is allogeneic. In other cases, cross-reactive structural epitopes on foreign MHC may bind TCR irrespective of the peptides it contains. The high numbers of MHC–TCR interactions between an APC and T cell are able to activate the T cells even if the binding affinity is low. Cross-reactivity between self- and non-self–MHC is much more common than cross-reactivity between any two random peptides, so the allogeneic response involves a higher proportion of T cells than the peptide-specific response.

Topic Test 4: Alloreactivity of T Lymphocytes

True/False

1. T cells most likely to react against allogeneic kidney cells are $CD8^+$ T_C.
2. Fewer T cells are activated by peptide–allogeneic MHC than by peptide–self-MHC because they are positively selected to recognize self-MHC in the thymus.

Multiple Choice

3. T cells that react against an allogeneic transplant
   a. Are gamma delta T cells, which were not educated to be restricted to self-MHC.
   b. Bind peptide + allogeneic MHC tightly enough to be activated.
   c. Bind native peptide in the absence of MHC.
   d. Escaped from the thymus without undergoing positive selection.
   e. Will be killed by this encounter.
4. Helper T cells are most likely to divide in response to
   a. Allogeneic hepatocytes (liver cells).
   b. Herpes virus peptides presented on syngeneic epithelial cells.
   c. Herpes virus peptides presented on syngeneic dendritic cells.
   d. Tissue macrophages in an allogeneic kidney.
   e. None of the above will stimulate T_H cell proliferation.

Short Answer

5. Why do so many T cells respond to foreign MHC?
6. What is the clinical significance of alloreactivity?

Topic Test 4: Answers

1. **True.** Allogeneic kidney cells have membrane MHC class I that can activate $CD8^+$ cytotoxic T cells.
2. **False.** The allogeneic response involves a higher percentage of T cells than the peptide-specific response.
3. **b.** Gamma delta TCR is not selected for MHC binding, so gamma delta T cells are unlikely to recognize allogeneic MHC. Binding to foreign MHC by mature T cells probably does not result in their death, although some may be inactivated; cytotoxic T lymphocytes (CTL) that recognize foreign MHC kill the allogeneic cells.
4. **d.** T_H cells proliferate in response to peptide on class II. Macrophages with allogeneic class II stimulate a larger response than herpes virus presented on syngeneic class II. Allogeneic hepatocytes and herpes virus peptides presented on syngeneic epithelial cells, which have membrane class I but not class II, activate $CD8^+$ T cells.
5. So many T cells respond to foreign MHC because there is considerable cross-reactivity between allogeneic MHC molecules.
6. Allogeneic responses are responsible for transplant rejection.

CLINICAL CORRELATION: SUPERANTIGENS

Toxic shock syndrome is caused by toxic shock syndrome toxin (TSST), an exotoxin produced by *Staphylococcus aureus*. TSST is a superantigen; it binds to the outer surface of TCR Vβ and class II without being processed. Its binding specificity for Vβ depends on the Vβ gene segment used, not on CDR3, so superantigens can bind up to 20% of all T cells. Superantigen binding activates T cells to produce massive quantities of cytokines, whose biological activities include blood vessel leakiness, hypovolemic shock, fever, widespread blood clotting, vomiting, and rashes.

DEMONSTRATION PROBLEM

To study MHC restriction of T cells, you make two clones of mouse $CD8^+$ T_C cells. The first is specific for peptide A presented on $H\text{-}2^k$ and the second is specific for peptide B presented on $H\text{-}2^d$. You then transfect cDNA for TCR from the cells of the first clone into cells from the second clone; the transfected T cells express both TCR molecules on each cell. Which of the following target cells would be killed by the transfected T cell clone and why?

TARGET CELLS	PEPTIDE
1. H-2^d	A
2. H-2^d	B
3. H-2^k	A
4. H-2^k	B
5. Congenic H-2^k background with K^d and D^d regions	A
6. Congenic H-2^k background with IA^d and IE^d regions	A

SOLUTION

Each transfected T cell makes two TCR, one specific for peptide A on H-2^k and one specific for peptide B on H-2^d. Because the TCR binds both peptide and MHC, only the proper peptide–MHC combination will be recognized, so targets 2 and 3 will be killed and targets 1 and 4 will not. Because the T cells are $CD8^+$, they see peptide on class I, so target 6 with the appropriate class I region for peptide A will be killed and target 5 with the inappropriate class I region will not be killed.

Chapter Test

True/False

1. Cells that begin somatic recombination of beta chain express neither CD4 nor CD8.
2. A superantigen is a commonly shared epitope found on many pathogens.
3. T cells that bind any self-peptides are negatively selected in the thymus.
4. MHC restriction of T cells has evolved to protect us from foreign tissue antigens.
5. A person deficient in membrane expression of HLA-DR and HLA-DP would probably have lower than normal numbers of $CD4^+$ T cells.

Multiple Choice

6. TCR most closely resembles Ig
   a. Alpha chain.
   b. Fab fragment.
   c. J chain.
   d. L chain.
   e. Membrane IgM.
7. Antigen receptors on both B cells and T cells are *not*
   a. Associated with signal transduction molecules in the membrane.
   b. Generated by somatic recombination during lymphocyte development.
   c. Members of the Ig gene superfamily.
   d. MHC-restricted in their ability to bind antigen.
   e. Specific for a single antigen epitope.
8. T-cell development in the thymus
   a. Begins with positive selection of double-negative progenitor cells.
   b. Involves both epithelial and marrow-derived APC.

c. Is successfully completed by 50% of all lymphoid progenitors entering the thymus.
d. Occurs primarily in the medulla.
e. Requires expression of RAG-1 and RAG-2 but not TdT.

9. Pre-Tα chain is required for
   a. Membrane expression of rearranged beta chain on double-positive T cells.
   b. Negative selection.
   c. Positive selection.
   d. The alpha chain of the CD3 complex.
   e. The alpha chain of the interleukin-2 receptor expressed early in T-cell development.

10. T-cell receptors on mature T cells
    a. Are both alpha beta and gamma delta, expressed simultaneously using alternative mRNA splicing.
    b. Are covalently bonded with either CD4 or CD8.
    c. Have been selected for their ability to recognize common pathogen epitopes on MHC.
    d. Recognize unprocessed allogeneic MHC directly.
    e. Rely on CD3 for increased avidity of TCR binding to APC.

Short Answer

11. Why do such large numbers of thymocytes fail positive selection?

12. Which thymus cells determine "self" for negative selection?

13. In a radiation chimera, why must donor bone marrow cells and recipient cells share at least one MHC antigen for normal T-cell responses to occur?

14. What is the B-cell equivalent of pre-Tα?

Essay

15. Why do you think antigen-driven TCR mutation outside the thymus would be discouraged in T cells?

Chapter Test: Answers

1. **T** 2. **F** 3. **F** 4. **F** 5. **T** 6. **a** 7. **d** 8. **b** 9. **a** 10. **d**

11. Most thymocytes fail positive selection because the random combination of V, D, and J segments and of alpha and beta chains rarely produces a TCR that binds with sufficient affinity to MHC class I or class II.

12. Bone marrow-derived macrophages and dendritic cells presenting self-peptides on class I and class II determine "self" for negative selection.

13. T cells being positively selected on recipient thymic epithelial cells and negatively selected on donor macrophages and dendritic cells must see the same MHC.

14. The B-cell equivalent of pre-Tα is λ5.

15. Antigen-driven TCR mutation outside the thymus is potentially dangerous for two reasons. First, the TCR might mutate so that the cell no longer recognized self-MHC, in which case it could not respond to antigen and would be useless. Second, the TCR might mutate to bind tightly to self-peptides on self-MHC, in which case the T cell would be autoreactive and dangerous. Because there is no positive or negative selection in the periphery, T cells undergoing these mutations could not be eliminated.

Check Your Performance:

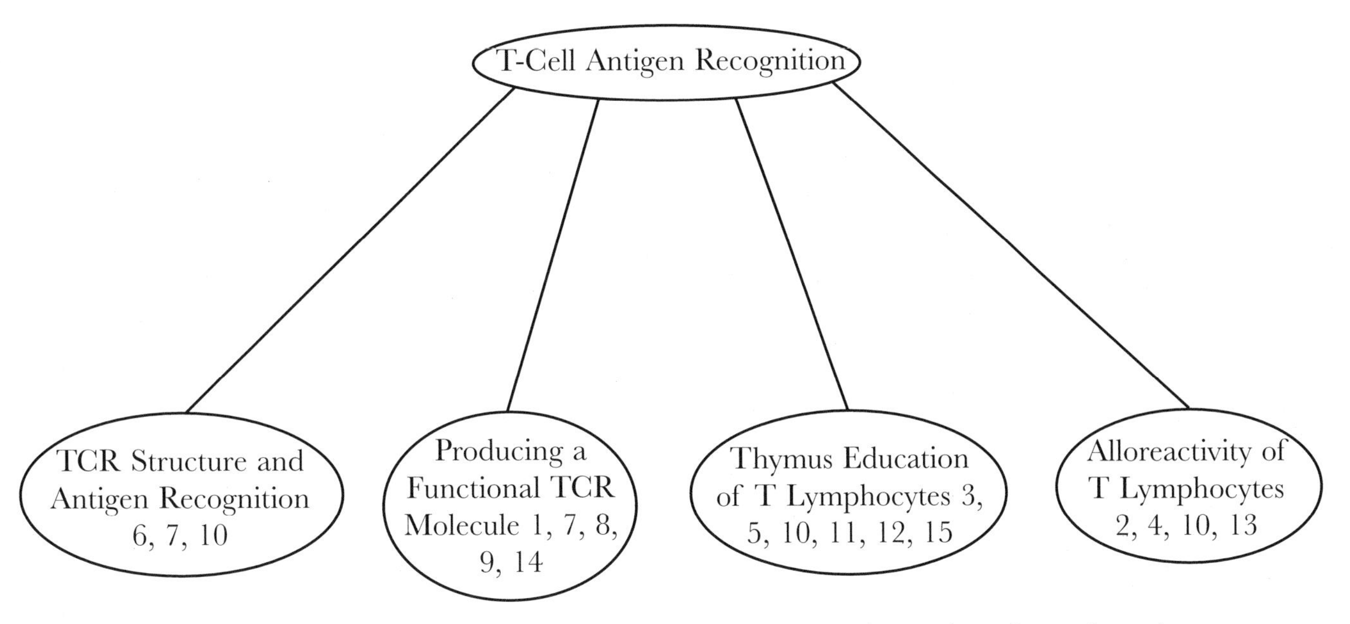

Check your understanding of this chapter by noting the number of questions for each topic you missed on the chapter test.

Midterm Exam

True/False

1. Immunity produced by vaccinating with killed virus is passive humoral immunity.
2. The antigen receptors on B cells (BCR) are IgM and IgG.
3. Blood typing is usually done using anti-A and anti-B antibodies to agglutinate red blood cells that have A, B, or both A and B antigens on their membranes.
4. Mouse anti-human IgG Fab is an anti-allotype antiserum.
5. The antibody isotype most protective against respiratory pathogens, because of its ability to block their binding to the respiratory epithelial cells, is secretory IgA.
6. Peptide binds to the alpha2 and beta2 domains of the major histocompatibility complex (MHC) class II proteins.
7. Human leukocyte antigen (HLA) DM prevents binding of endogenous peptides to MHC class II in the endosome.
8. Macrophages stop antigen processing when all their MHC molecules are full of peptide.
9. Antigen-specific T lymphocytes develop from stem cells in the bone marrow and mature in the thymus in response to contact with specific antigen.
10. Positive selection is responsible for MHC restriction of T-cell responses.

Multiple Choice

11. Secondary lymphoid organs
    a. Are the primary site of RAG-1 and RAG-2 expression.
    b. Bring antigen into contact with antigen-presenting cells (APC) and mature lymphocytes.
    c. Expose T cells to self-antigen so they can be negatively selected.
    d. Provide the microenvironment for differentiation of mature T and B lymphocytes.
    e. Store long-lived effector cells after an immune response.
12. The molecule that would probably be *least* immunogenic in humans is
    a. A complex of tetanus toxoid (a protein) and streptococcal polysaccharide.
    b. Antibody against chickenpox virus from another human.
    c. Antibody against snake venom produced in a horse.
    d. Bacterial lipopolysaccharide, a B-cell mitogen.
    e. Penicillin bound to human serum proteins.
13. Antibody Fab fragments do *not* contain
    a. Antigen-binding sites.
    b. CDR.
    c. Complement-binding sites.
    d. Idiotypic determinants.
    e. Light chain variable regions
14. IgG is *not*
    a. Able to transfer immunity across the placenta from mother to fetus.

b. Composed of two gamma chains and two light chains.
c. Part of the Ig superfamily.
d. The largest Ig.
e. The predominant antibody isotype in blood, lymph fluid, and peritoneal fluid.

15. The enormous diversity of antibody molecules is due to
a. Alternative splicing of Ig mRNA molecules.
b. Co-dominant expression of many Ig molecules on individual B cells.
c. Gene segment rearrangement until an antibody that binds antigen is produced.
d. The large number of possible V, D, and J gene combinations that can be made.
e. The remarkable cross-reactivity of antibody molecules.

16. As a B cell develops, productive rearrangement of H chain gene occurs but kappa chain gene rearrangement on both chromosomes is nonproductive. The B cell will then
a. Begin lambda chain gene segment rearrangement.
b. Begin rearranging its surrogate L chain genes.
c. Die.
d. Undergo isotype switching.
e. Undergo somatic mutation to produce a functional kappa chain gene.

17. Expression of class I molecules is *not* associated with
a. Antigen presentation to cytotoxic T lymphocytes (T_C).
b. Graft rejection.
c. Increased risk of certain autoimmune diseases.
d. Lysis of virus-infected cells.
e. Stimulation of macrophage cytotoxicity by helper T cells.

18. Exogenous antigen is processed
a. By the endosomal processing pathway.
b. In every nucleated cell.
c. In the presence of HLA-DM.
d. In proteasomes.
e. After co-stimulation by APC.

19. The role of CD3 in T-cell–APC interactions is to
a. Attach T-cell receptor (TCR) to the T-cell membrane.
b. Be a cytokine receptor.
c. Convert a membrane signal into a cytoplasmic signal.
d. Signal the APC to process more antigen.
e. Stabilize the TCR–peptide–MHC complex.

20. Development of $CD8^+$ T cells could occur normally in the absence of
a. HLA-DR genes.
b. RAG-1 and RAG-2.
c. Self-peptides.
d. Tapasin and calnexin.
e. TAP-1 and TAP-2.

Short Answer

21. Describe two similarities and two differences between Ig and MHC proteins.

22. What is humanized antibody and for what purpose(s) is it used?

23. Describe the differences in antigen recognition by B cells and T_C cells. Include in your answer the differences between antigen epitopes seen by B cells and T_C cells, molecules involved in antigen recognition on each cell type, and body locations where each cell type is likely to encounter antigen.

24. You are studying B-cell development in the bone marrow. You have available normal mice, fluorescein isothiocyanate (FITC)-labeled anti-mu chain monoclonal antibodies, and rhodamine-labeled anti-delta chain monoclonal antibodies. (FITC and rhodamine emit green and red fluorescence, respectively.) After treating bone marrow cells with both antibodies, you subject your cells to flow cytometry analysis. Describe the staining results you would expect for mature B cells, immature B cells, and non-B cells.

Essay

25. Describe internalization, processing, and presentation of an endogenous antigen by a B cell.

Midterm Exam: Answers

1. **F** 2. **F** 3. **T** 4. **F** 5. **T** 6. **F** 7. **F** 8. **F** 9. **F** 10. **T** 11. **b** 12. **b**

13. **c** 14. **d** 15. **d** 16. **d** 17. **e** 18. **a** 19. **c** 20. **a**

21. Both Ig and MHC proteins belong to the Ig superfamily and have disulfide bond-linked domains. Both are membrane proteins on B cells, composed of two different polypeptide chains, and have variable antigen-binding regions and constant membrane proximal regions of several classes (isotypes). Ig has four chains whereas MHC has two, Ig variable regions are much more variable, and MHC variability is greater in the population than in the individual. Ig is secreted and membrane bound; MHC is expressed on more cell types.

22. Humanized antibody has mouse monoclonal CDR and human monoclonal framework and Fc regions. It is used for cancer diagnosis and therapy in humans and is less immunogenic than mouse monoclonal antibody.

23. B cells bind both linear and assembled epitopes. T cells respond primarily to processed and presented peptides, TCR affinity for antigen is lower than that of BCR, and TCR binds MHC. TCR function requires CD8 on T_C; both B and T cells require signal transduction molecules (Ig alpha/Ig beta or CD3) and adhesion molecules for their activation. BCR specificity changes due to somatic hypermutation, and secreted antibodies have a variety of biological effector functions. Both B and T_C cells are usually activated in the secondary lymphoid organs but may also bind antigen in the periphery.

24. Mature B cells stain with both FITC-anti-mu and rhodamine anti-delta. Immature B cells stain only with anti-mu, and non-B cells do not stain with either antibody.

25. Endogenous antigen, for example, proteins from an infecting virus, is synthesized on ribosomes in the B-cell cytoplasm. To be presented, these proteins are degraded to peptides in proteasomes. Two proteases encoded in MHC class II (LMP2 and LMP7) and a third subunit not encoded in MHC are produced in response to interferon, which

is synthesized in response to virus infection. These inducible proteases replace constitutive proteases in the proteasome and produce peptides with basic and hydrophobic carboxyl terminal residues preferred as anchor residues in class I peptide binding sites and for transport to the endoplasmic reticulum (ER). TAP-1 and TAP-2 are present in the ER membrane with ATP-binding domains on the cytosolic side and hydrophobic transmembrane domains spanning the ER membrane. The TAP-1–TAP-2 complex transports cytosolic peptides into the lumen of the ER with the expenditure of ATP. Newly synthesized partly folded class I alpha chain binds calnexin in the lumen of the ER. When beta$_2$-microglobulin binds to class I, calnexin dissociates and class I alpha-beta$_2$-microglobulin forms a complex with calreticulin, tapasin, and TAP transporter. When class I binds a peptide, it is released from the TAP transporter and the class I–peptide complex is transported through the Golgi to the plasma membrane.

Unit III:
Immune Effector Mechanisms

Cytokines

Cytokines are the messenger proteins through which cells in the immune system communicate. They are secreted in response to antigen or cytokine stimuli and generally act over a short distance with a short half-life. Cytokines bind specific receptors whose expression and affinity are key factors in regulating cellular responses.

ESSENTIAL BACKGROUND

- **Immune system cells and molecules (Chapter 1)**
- **Inducing, detecting, and investigating immunity (Chapter 2)**

TOPIC 1: GENERAL PROPERTIES OF CYTOKINES

KEY POINTS

✓ *What kind of molecules are cytokines?*

✓ *How do cytokines transmit their information to cells?*

✓ *What is meant by cytokine pleiotropy and redundancy?*

Cytokines are small secreted proteins that mediate and regulate immunity, inflammation, and hematopoiesis. They are produced de novo in response to immune stimuli and act over short distances, short time spans, and very low concentrations. They bind specific membrane receptors, which then signal the cell via second messengers, often tyrosine kinases, to alter behavior (gene expression). **Lymphokines** are cytokines made by lymphocytes, **monokines** are cytokines made by monocytes, **chemokines** are chemotactic cytokines, and **interleukins** (ILs) are cytokines made by one leukocyte and acting on other leukocytes.

Cytokines act on the cells that secrete them (**autocrine** action), on nearby cells (**paracrine** action), or on distant cells (**endocrine** action). Several cell types secrete the same cytokine, and a single cytokine acts on several different cell types (**pleiotropy**) (**Table 7.1**). Similar functions can be stimulated by different cytokines (**redundancy**). Cytokines are often produced in a **cascade**, as one cytokine stimulates its target cells to make additional cytokines. Cytokines also act **synergistically** (two or more cytokines acting together) or **antagonistically** (cytokines causing opposing activities).

Table 7.1 Selected Immune Cytokines and Their Activities*

CYTOKINE	PRODUCING CELL	TARGET CELL	FUNCTION
GM-CSF	T_H	Progenitor cells	Growth and differentiation of monocytes and dendritic cells
IL-1 alpha IL-1 beta	Monocytes, M_ϕ, B cells, dendritic cells	T_H cells B cells NK cells Various	Co-stimulation Maturation and proliferation Activation Inflammation, acute phase response, fever
IL-2	T_H1 cells	Ag-primed T and B cells, NK cells	Growth, proliferation Activation
IL-3	T_H cells, NK cells	Stem cells Mast cells	Growth and differentiation Growth and histamine release
IL-4	T_H2 cells	Ag-primed B cells Macrophages T cells	Proliferation and differentiation Ig class switch to IgG_1 and IgE MHC class II expression Proliferation
IL-5	T_H2 cells	Activated B cells	Proliferation and differentiation Ig class switch to IgA
IL-6	Monocytes, M_ϕ, T_H2 cells, stromal cells	Activated B cells Plasma cells Stem cells Various	Differentiation into plasma cells Antibody secretion Differentiation Acute phase response
IL-7	Marrow and thymic stroma	Stem cells	Differentiation into progenitor B and T cells
IL-8	M_ϕ, endothelial cells	Neutrophils	Chemokine
IL-10	T_H2 cells	M_ϕ B cells	*Cytokine production* Activation
IL-12	M_ϕ, B cells	Activated T_C cells NK cells	Differentiation into CTL (with IL-2) Activation
IFN alpha	Leukocytes	Various	*Viral replication*, MHC I expression
IFN beta	Fibroblasts	Various	*Viral replication*, MHC I expression
IFN gamma	T_H1 cells, T_C, NK cells	Various M_ϕ Activated B cells T_H2 cells M_ϕ	*Viral replication* MHC expression Ig class switch to IgG_{2a} *Proliferation* DTH activity
MIP-1 alpha	M_ϕ	Monocytes, T cells	Chemotaxis
MIP-1 beta	Lymphocytes	Monocytes, T cells	Chemotaxis
TGF beta	T cells, monocytes	Monocytes and M_ϕ Activated M_ϕ Various Activated B cells	Chemotaxis IL-1 synthesis *Proliferation* IgA synthesis
TNF alpha	M_ϕ, mast cells, NK cells	M_ϕ Tumor cells	CAM and cytokine expression Cell death
TNF beta	T_H1 and T_C cells	Phagocytes Tumor cells	Phagocytosis, NO production Cell death

* Italicized activities are inhibited.
M_ϕ, macrophage; Ig, immunoglobulin; MHC, major histocompatibility class; TGF, transforming growth factor; CTL, cytotoxic T lymphocyte; DTH, delayed type hypersensitivity; CAM, cell adhesion molecule; NO, nitric oxide.

Topic Test 1: General Properties of Cytokines

True/False

1. A cytokine is an example of a lymphokine.
2. Cytokines bind to nonspecific receptors on white blood cells.

Multiple Choice

3. Cytokines may exhibit ____________ action, signaling ____________.
   a. Antagonistic; other cytokines.
   b. Autocrine; different target cells to produce different effects.
   c. Endocrine; distant target cells.
   d. Paracrine; receptors on the cell that secreted the cytokines.
   e. Synergistic; two different cell types.
4. Cytokines are not
   a. Antigen specific.
   b. Capable of activating more than one cell type.
   c. Made by lymphocytes.
   d. Small protein molecules.
   e. Synthesized de novo in response to antigen or other cytokines.

Short Answer

5. What is a chemokine?
6. What factors limit the activity of a cytokine?

Topic Test 1: Answers

1. **False.** A lymphokine is an example of a cytokine.
2. **False.** Cytokine receptors are cytokine specific (but not antigen specific).
3. **c.** Cytokines may act on the cytokine-producing cell (autocrine activity), a nearby cell (paracrine), or a distant cell (endocrine).
4. **a.** The same cytokines are made in response to many different antigens.
5. A chemokine is a chemoattractant cytokine.
6. Cytokine activity is limited by its synthesis, short half-life, distance to the target cell, and availability of specific receptors on target cells.

TOPIC 2: CYTOKINE FUNCTIONS

KEY POINTS

✓ *How are functions of specific cytokines identified?*

✓ *What are the activities of immune cytokines?*

✓ *What are the activities of T_H1 and T_H2 cells?*

Cytokines are made by many cell populations, but the predominant producers are helper T cells (T_H) and macrophages. Table 7.1 lists the sources and activities of many immune cytokines. The largest group of cytokines stimulates immune cell proliferation and differentiation, including IL-1, which activates T cells; IL-2, which stimulates proliferation of antigen-activated T and B cells; IL-4, IL-5, and IL-6, which stimulate proliferation and differentiation of B cells; interferon gamma (IFN gamma), which activates macrophages; and IL-3 and granulocyte-monocyte colony-stimulating factor (GM-CSF), which stimulate hematopoiesis. Other groups of cytokines include IFNs that inhibit virus replication in infected cells and chemokines that attract leukocytes to infection sites. Chemokines have conserved cysteine residues that allow them to be assigned to four groups. The groups, with representative chemokines, are C-C chemokines (RANTES, MCP-1, MIP-1 alpha, and MIP-1 beta*), C-X-C chemokines (IL-8), C chemokines (lymphotactin), and CXXXC chemokines (fractalkine). Some cytokines are predominantly inhibitory; for example, IL-10 and IL-13 inhibit inflammatory cytokine production by macrophages.

T_H have two important functions: to stimulate cellular immunity and inflammation and to stimulate B cells to produce antibody. Two functionally distinct subsets of T cells secrete cytokines that promote these different activities. T_H1 cells produce IL-2, IFN gamma, and tumor necrosis factor beta (TNF beta), which activate cytotoxic T lymphocytes (T_C) and macrophages to stimulate cellular immunity and inflammation. T_H1 cells also secrete IL-3 and GM-CSF. T_H2 cells secrete IL-4, IL-5, and IL-6, which stimulate antibody production by B cells. T cells are initially activated as T_0 cells, which produce IL-2, IL-4, and IFN gamma. The cytokine environment then influences differentiation into T_H1 or T_H2 cells. IL-4 stimulates T_H2 activity and suppresses T_H1 activity, whereas IL-12 promotes T_H1 activities. T_H1 and T_H2 cytokines are antagonistic. T_H1 cytokine IFN gamma inhibits proliferation of T_H2 cells and stimulates B cells to secrete IgG_{2a}. T_H2 cytokine IL-10 inhibits T_H1 secretion of IFN gamma and suppresses class II expression and production of inflammatory cytokines by macrophages. The balance between T_H1 and T_H2 activity steers the immune response toward cellular or humoral immunity.

Topic Test 2: Cytokine Functions

True/False

1. T_H1 and T_H2 cells develop in separate areas of the thymus where the cytokine environments are different.
2. IFNs inhibit virus replication by infected cells.

Multiple Choice

3. Characterization of cytokine activities is *not* made more difficult by their
   a. Gene structure.
   b. Pleiotropism.
   c. Redundancy.
   d. Secretion.
   e. Short half-lives.

* RANTES = regulation upon activation normal T cell–expressed and secreted; MCP = macrophage chemoattractant and activating factor; MIP = macrophage inflammatory protein.

4. T_H1 cells secrete cytokines that
   a. Activate cytotoxic T cells.
   b. Increase susceptibility to allergic reaction.
   c. Inhibit macrophage secretion of inflammatory cytokines.
   d. Stimulate B cells to secrete IgG_1.
   e. Stimulate proliferation of T_H2 cells.

Short Answer

5. Why are knock-out mice important for characterizing cytokines?
6. What is meant by the antagonistic activities of T_H1 and T_H2 cytokines?

Topic Test 2: Answers

1. **False.** T_H1 and T_H2 subsets develop in areas of the peripheral lymphoid organs and tissues where the cytokine environments are different.
2. **True.** IFN alpha and IFN beta are made by host cells and bind to virus-infected cells to inhibit virus replication. IFN gamma is made by T_H1 cells and stimulates cellular immunity.
3. **a.** Characterization of cytokine activities often begins by looking for genes homologous with other known cytokine genes.
4. **a.** T_H1 cells secrete cytokines that activate cytotoxic T cells and macrophages and that inhibit antibody production and proliferation of T_H2 cells.
5. Knock-out mice, in which the genes for a single cytokine are deleted or inactivated, allow immunologists to do in vivo studies of individual cytokines under physiological conditions.
6. T_H1 and T_H2 cytokines behave antagonistically by inhibiting each other's production and the proliferation of the opposing T-cell subset.

TOPIC 3: CYTOKINE RECEPTORS AND ANTAGONISTS

KEY POINTS

✓ *What are the structures of cytokine receptors?*

✓ *How does cytokine binding to its receptors alter target cell function?*

✓ *How are cytokine activities targeted to antigen-specific lymphocytes?*

✓ *What is the importance of cytokine antagonists?*

Cytokines act on their target cells by binding specific membrane receptors. The receptors and their corresponding cytokines have been divided into several families based on their structure and activities. Hematopoietin family receptors include receptors for IL-2 through IL-7 and GM-CSF. IFN family receptors include receptors for IFN alpha, IFN beta, and IFN gamma. TNF family receptors bind soluble TNF alpha and TNF beta and membrane-bound CD40 (important for B-cell and macrophage activation) and Fas (which signals the cell to undergo apoptosis). Chemokine family includes receptors for IL-8, MIP-1, and RANTES; they allow human immunodeficiency virus to enter host cells and are responsible for the virus preferentially infecting either T cells or macrophages.

Hematopoietin cytokine receptors are the best characterized. They generally have two subunits, one cytokine specific and one signal transducing. An example is the GM-CSF subfamily, where a unique alpha subunit specifically binds GM-CSF, IL-3, or IL-5 with low affinity and a shared beta subunit signal transducer also increases cytokine-binding affinity. Cytokine binding promotes dimerization of alpha and beta subunits, which then associate with cytoplasmic tyrosine kinases that phosphorylate proteins that activate mRNA transcription. GM-CSF, IL-3, and IL-5 cause eosinophil proliferation and basophil degranulation and phosphorylate the same cytoplasmic protein. GM-CSF and IL-3 act on hematopoietic stem cells and progenitor cells and activate monocytes. Antagonistic GM-CSF and IL-3 activities are explained by their competition for limited amounts of beta subunit.

The IL-2R subfamily of receptors for IL-2, IL-4, IL-7, IL-9, and IL-15 also share a signal-transducing gamma chain. Each has a unique cytokine-specific alpha chain; IL-2 and IL-15 are trimers and share an IL-2R beta chain. Monomeric IL-2R alpha has low affinity for IL-2, dimeric IL-2R beta gamma has intermediate affinity, and trimeric IL-2R alpha beta gamma is high-affinity. Alpha chain (also called TAC) is expressed by activated but not resting T cells, which along with natural killer (NK) cells constitutively express low numbers of IL-2R beta gamma. Antigen activation stimulates T-cell expression of high affinity IL-2R trimers and IL-2 secretion, allowing autocrine stimulation of T-cell proliferation in an antigen-specific manner. Antigen specificity of the immune response is also maintained by the close proximity of antigen-presenting B cells and macrophages with their helper T cells. **X-linked severe combined immunodeficiency** (**X-SCID**) is caused by a defect in IL-2R family gamma chain, which results in loss of activity from this family of cytokines.

Cytokine activity is blocked by molecules that bind cytokines or their receptors. IL-1 has a specific antagonist, which blocks binding of IL-1 alpha and IL-1 beta to IL-1R. During immune responses, fragments of membrane receptors may be released and compete for cytokine. Vaccinia virus makes soluble molecules that bind IFN gamma, whereas Epstein-Barr virus makes a molecule homologous to IL-10 that suppresses host immune function.

Topic Test 3: Cytokine Receptors and Antagonists

True/False

1. Some members of the TNF receptor family bind lymphocyte membrane proteins.
2. Inhibitory cytokines act by blocking stimulatory cytokine binding to a shared receptor.

Multiple Choice

3. IL-2 receptor
   a. Carries IL-2 into the cytoplasm, where it phosphorylates proteins that increase mRNA transcription for proteins required for T-cell proliferation.

b. Consists of a high-affinity IL-2 binding chain and two signal-transducing units that do not bind IL-2.
c. Has unique cytokine binding and signal transducing chains.
d. Is present on resting T cells in high numbers.
e. Shares a signal-transducing subunit with IL-4R.

4. Members of a cytokine receptor family
   a. All bind the same cytokines.
   b. Are grouped together because they share antigen specificity.
   c. Are often found on the same cells.
   d. Are similar in protein structure and sometimes in regions of amino acid sequence.
   e. Are specific for cytokines produced by a single cell type.

Short Answer

5. Compare the structures and relative affinities of IL-2R on resting and antigen-stimulated T cells.

6. Considering the redundancy of cytokines, where several have the same function, how can a deficiency in one cytokine receptor subunit result in an immunodeficiency in both humoral and cellular immune systems (X-SCID)?

Topic Test 3: Answers

1. **True.** Examples are CD40 and Fas.

2. **False.** Antagonistic cytokines may compete for a shared signal transducing subunit, but some inhibitory cytokines act via distinct receptors to block proliferation or cytokine secretion of particular target cells.

3. **e.** IL-2R and IL-4R share a signal transducing gamma chain, which signals changes in gene expression via cytoplasmic second messengers. IL-2R beta chain is also a signal transducer but binds IL-2 to increase binding affinity.

4. **d.** Cytokine receptor families are formed based on their structure. Members of some families share cytokine activities (like the chemokine and IFN families) and signal transducing subunits, but they bind distinct cytokines and may be found on many different cell types.

5. Resting T cells express low numbers of IL-2R beta gamma, which has an intermediate affinity for IL-2. Antigen activation stimulates T-cell expression of higher numbers of high-affinity IL-2R alpha beta gamma trimers.

6. The receptor subunit that is deficient in X-SCID is the gamma chain signal-transducing subunit shared by receptors for IL-2, IL-4, IL-7, IL-9, and IL-15. These cytokines stimulate stem cell differentiation into progenitor B and T cells (IL-7), T-cell proliferation (IL-2, IL-9, and IL-15), and B-cell proliferation and differentiation (IL-4 and IL-15).

CLINICAL CORRELATION: LEPROSY (HANSEN'S DISEASE)

Leprosy is caused by infection with *Mycobacterium leprae*, a slow-growing organism that survives and replicates in macrophage phagosomes. Leprosy is endemic in parts of California, Florida, Hawaii, Louisiana, and Texas and in many countries in Africa and Asia. Two forms of leprosy are associated with different T_H responses. In tuberculoid leprosy, a predominantly T_H1 response is made to *M. leprae*. IL-2, IFN gamma, and TNF beta are produced, and few bacteria are detectable in macrophages. The disease causes local inflammation, granulomas, and peripheral nerve damage but spreads slowly. T-cell responses to other antigens and serum antibody levels are normal; a specific immune response to *M. leprae* is detectable. Treatment with antibiotics is usually curative. Lepromatous leprosy is characterized by high numbers of *M. leprae* in macrophages and disseminated tissue, bone, cartilage, and nerve damage. T-cell responses to other antigens are suppressed, and IL-4, IL-5, and IL-6 (T_H2 cytokines) predominate. Serum antibody is abnormally high; no specific response to *M. leprae* is detectable.

DEMONSTRATION PROBLEM

Compare and contrast the consequences of a deletion in the gene encoding IL-3 and in the gene encoding the beta subunit of IL-3 receptor.

SOLUTION

A deletion in the IL-3 gene results in a lack of IL-3 production by T_H, NK cells, and mast cells. IL-3 stimulates growth and differentiation of stem cells and growth and histamine release by mast cells, so hematopoiesis and inflammation would be reduced. However, other cytokines such as GM-CSF, IL-5, IL-6, IL-7, TNF alpha, and TNF beta share the ability of IL-3 to stimulate hematopoiesis and inflammation. Deletion in the beta subunit of the IL-3 receptor also results in lack of IL-3 activity, because the beta subunit is required for signal transduction. However, the beta subunit is shared by receptors for GM-CSF and IL-5, so these cytokines could not replace IL-3 function and the deficit in hematopoiesis and inflammatory cells would be more severe.

Chapter Test

True/False

1. Chemokines stimulate leukocyte migration to an infection site.
2. The ability of an intracellular pathogen to induce a strong T_H2 response may increase the survival of the pathogen.
3. Redundancy means that several cytokines may have the same effect on the cells they bind.
4. The prolonged half-life of secreted cytokines contributes to immune memory.
5. IL-1 antagonist is a molecule that blocks IL-1 binding to its specific receptor.

Multiple Choice

6. The ability of a cytokine to change gene expression in the target cell is influenced by all of the following *except*
   a. Presence of high-affinity receptors on the target cell.
   b. Presence of soluble cytokine receptors.
   c. Proximity of the producing and target cells.
   d. Rate of transport of cytokine–receptor complexes into the cytoplasm.
   e. Simultaneous production of another cytokine whose receptor uses the same signal transducing subunit.

7. Cytokines are *not*
   a. Able to inhibit the function of other cytokines.
   b. Able to stimulate the synthesis of other cytokines
   c. Produced by more than one cell type.
   d. Small protein molecules
   e. Stored in the cell for quick release.

8. The IL-2R subfamily contains receptors for IL-2, IL-4, IL-7, IL-9, and IL-15. This group of cytokine receptors
   a. Binds all five cytokines to promote synergistic action on target cells.
   b. Binds cytokines that are produced by the same cell.
   c. Each has a unique high-affinity cytokine-specific alpha chain.
   d. Shifts the immune response toward cellular immunity.
   e. Shares a signal-transducing gamma chain.

9. An antagonist for cytokine X may *not* be
   a. Cytokine A, competing for a shared receptor subunit.
   b. Cytokine B, which acts synergistically with cytokine X.
   c. Cytokine C, which inhibits the activation of the cell that produces cytokine X.
   d. Made by microorganisms.
   e. Soluble cytokine X receptors.

10. A knock-out mouse for a particular cytokine allows immunologists to characterize cytokine function
    a. By doing a dose-response study with competing cytokines.
    b. In the absence of all other cytokines.
    c. On all cell types simultaneously.
    d. Under controlled conditions of local cytokine concentrations.
    e. With defined cell populations.

Short Answer

11. How do cytokines transmit their information to cells?
12. What immune effector functions are stimulated by T_H1 and T_H2 cells?
13. Describe the IL-2 receptor.
14. After a booster for measles virus, memory T_H cells are presented with antigen by measles-specific memory B cells and secrete IL-4 to stimulate B-cell proliferation and antibody synthesis. Although many B cells are present, the antibody response is measles-virus specific; why aren't other B cells stimulated by the IL-4?

Essay

15. Describe the two forms of leprosy and explain the significance of the T-cell cytokine secretion profiles in each form of the disease.

Chapter Test: Answers

1. **T** 2. **T** 3. **T** 4. **F** 5. **T** 6. **d** 7. **e** 8. **e** 9. **b** 10. **c**

11. Cytokines bind specific receptors that then associate with cytoplasmic second messengers, usually kinases.

12. T_H1 cytokines include IL-2 and IFN gamma, which activate T_C to kill virus-infected cells and macrophages to kill vesicular pathogens and secrete inflammatory cytokines. T_H2 cytokines include IL-4, IL-5, and IL-6 that stimulate B cells to divide and differentiate into antibody-secreting plasma cells. IL-4 is particularly important for IgE secretion and allergies.

13. IL-2R has a unique low-affinity cytokine-specific alpha chain, a signal-transducing beta chain that it shares with IL-5R, and a signal-transducing gamma chain that it shares with all members of the IL-2R family. Resting T cells and NK cells constitutively express low numbers of IL-2R beta gamma, which have intermediate affinity for IL-2. Antigen activation stimulates T-cell expression of alpha chain to produce high affinity IL-2R alpha beta gamma trimers.

14. Antigen and T_H binding stimulate measles-specific B cells to express more IL-4 receptors. In addition, close proximity to measles-specific T_H increases the effective concentration of IL-4 at the surface of the specific B cell.

15. Leprosy is caused by infection with *M. leprae*, which replicates in macrophage phagocytic vesicles. Tuberculoid leprosy occurs when the immune response is predominantly T_H1 cytokines IL-2, IFN gamma, and TNF beta. These cytokines activate macrophages to kill bacteria in their vesicles, so few bacteria are detectable in macrophages, and inflammation, granulomas, and peripheral nerve damage are localized. Lepromatous leprosy occurs when the immune response is predominantly T_H2 cytokines IL-4, IL-5, and IL-6. Macrophages are not activated to kill the vesicular bacteria, and high numbers of *M. leprae* are present. The infection spreads rapidly and causes disseminated tissue, bone, cartilage, and nerve damage. T-cell responses to *M. leprae* are suppressed because T_H2 cytokines inhibit proliferation of and cytokine production by T_H1 cells, and serum antibody is abnormally high as the immune system tries unsuccessfully to eliminate the pathogen with humoral immune mechanisms.

Check Your Performance:

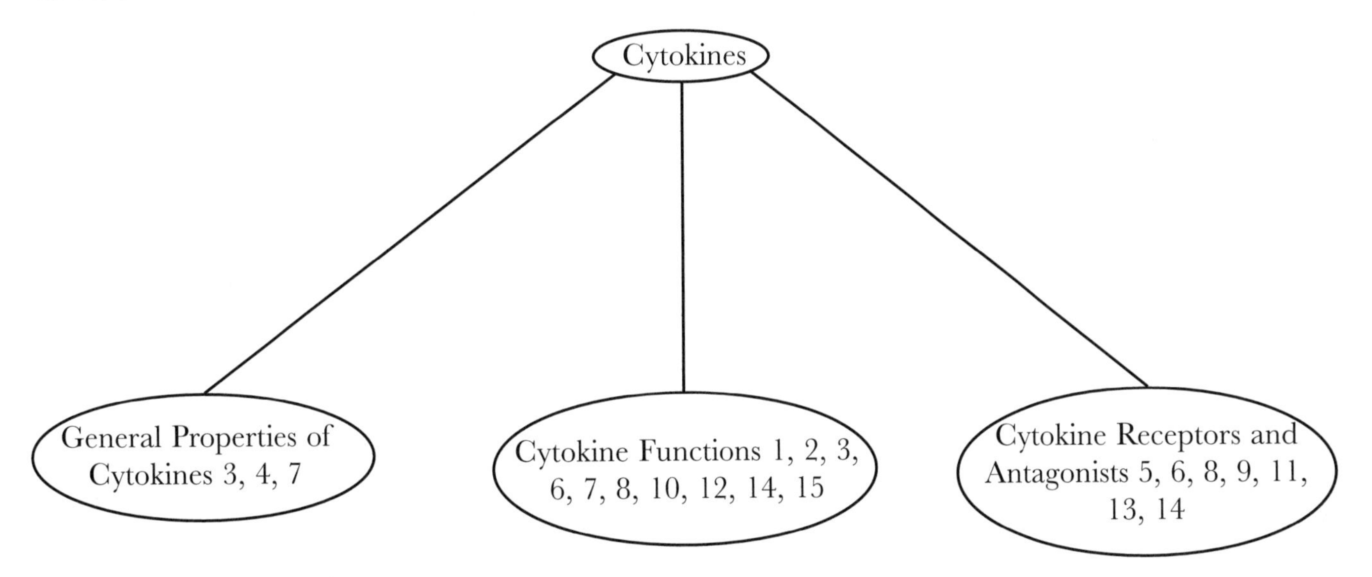

Check your understanding of this chapter by noting the number of questions for each topic you missed on the chapter test.

Complement

Complement is a group of serum proteins that work with or "complement" antibody activity to eliminate pathogens. Complement is not antigen specific, and activation occurs immediately in the presence of activating molecules, making it part of innate immunity. Antibody activates some complement proteins, so complement also participates in humoral immunity. Complement stimulates inflammation, facilitates antigen phagocytosis, and lyses some cells directly. Because it is such a powerful inflammatory agent, its activity is tightly regulated.

ESSENTIAL BACKGROUND

- **Innate and adaptive immunity (Chapter 1)**
- **Inducing, detecting, and investigating immunity (Chapter 2)**
- **Antibody structure and function (Chapter 3)**

TOPIC 1: FUNCTIONS OF COMPLEMENT

KEY POINTS

✓ *What molecules can activate the complement cascade?*

✓ *What effector functions result from complement activation?*

✓ *What effector cells have complement receptors?*

Complement proteins are produced constitutively by macrophages and hepatocytes and are present in the circulation as inactive molecules. Several complement proteins are proteolytic **proenzymes**; when activated, they cut peptide bonds in other complement proteins to activate their protease function in an enzymatic **cascade**. Because each protease activates many substrate molecules, the initial activation is rapidly amplified. Two complement cascades are activated by different molecules but share some constituents. The **classical complement pathway** activated by antigen-bound IgM and IgG was discovered first. However, **the alternative complement pathway** activated by bacterial cell surface molecules probably evolved first. Some molecules that activate the alternative complement cascade include lipopolysaccharide from gram-negative outer membranes, teichoic acid from gram-positive cell walls, zymosan from fungal and yeast cell walls, and parasite surface molecules.

Complement proteins are numbered in the order in which they were discovered, which is nearly the order in which they are activated. C1 is a complex of three molecules: C1q, C1r, and C1s.

C1r and C1s are proteases. During activation, some complement components are split into two pieces; the larger fragment "b" usually remains attached to the antigen and the smaller fragment "a" often diffuses away. Activated complement fragments with enzymatic activity are denoted by placing a bar over their numbers. Inactivated fragments are indicated by a small "i."

Complement activation stimulates several antimicrobial activities. **Opsonin** fragments adhere to microorganisms and promote leukocyte chemoattraction, antigen binding and phagocytosis, and activation of killing mechanisms. **Anaphylatoxins** promote inflammation by binding mast cell complement receptors and triggering release of histamine, which increases blood vessel permeability and smooth muscle contraction. **Membrane attack complex** (MAC) inserts into lipid membranes of bacteria or eukaryotic cells and causes osmotic lysis. Complement also coats viruses to block host cell binding and promotes immune complex removal. Many bacteria have mechanisms to evade complement-mediated damage.

Complement binds specific receptors on various cell types to mediate inflammatory and opsonic activities (**Table 8.1**). The best characterized is CR1, which binds opsonin fragments C3b and C4b and promotes phagocytosis and clearance of antigen–antibody complexes in combination with antibody binding to Fc receptor (FcR). Receptor for C1q also promotes immune complex binding to phagocytes. CR2 is part of the B-cell receptor complex; binding of antigen-complement complexes to CR2 increases B-cell antigen sensitivity by up to a thousandfold. Epstein-Barr virus (EBV) uses CR2 to enter B cells, causing infectious mononucleosis and occasionally transforming the B cell into a B-lymphoma cell. CR3 and CR4 are better known as integrin molecules MAC-1 and p150.95; they allow monocytes, macrophages, neutrophils, and dendritic cells to adhere to blood vessel walls and move into the tissues (**extravasate**) at the site of inflammation. Receptors for anaphylatoxins signal mast cell degranulation and smooth muscle contraction. C5a receptor signals macrophage phagocytosis of complement-coated antigen in the absence of IgG binding to FcR, which is important early in a primary response when no antibody or only IgM is present.

Table 8.1 Complement Receptors

RECEPTOR	LIGANDS	FUNCTIONS
CR1 (CD35)	C3b, C3bi, C4b, C4bi	Opsonization and antigen clearance (phagocytes) Antigen persistence (follicular dendritic cells) Clearance of immune complexes (red blood cell) Regulation of complement cascade
CR2 (CD21)	C3d, C3dg, C3bi (Epstein-Barr virus)	B-cell activation (B cells, FDC)
CR3 (MAC-1, CD11b/CD18)	C3bi	Adhesion, extravasation (monocytes, macrophages, neutrophils) Phagocytosis (macrophages, neutrophils)
CR4 (p150.95, CD11c/CD18)	C3bi	Adhesion, extravasation (macrophages, neutrophils, dendritic cells) Phagocytosis (macrophages, neutrophils)
C1qR	C1q	Immune complex binding to phagocytes (macrophages, neutrophils)
C5aR	C5a	Adherence, phagocytosis, CR1 and CR3 expression (macrophages, neutrophils)
C3aR, C4aR, C5aR	C3a, C4a, C5a	Degranulation (mast cells), contraction (smooth muscle)

FDC, follicular dendritic cell.

Topic Test 1: Functions of Complement

True/False

1. The classical complement cascade is activated by IgM or IgG bound to antigen.
2. Some complement receptors are also adhesion molecules that allow leukocytes to move from the circulation into the tissues.

Multiple Choice

3. Complement
   a. Is a group of active proteolytic enzymes found in serum.
   b. Is secreted by macrophages and hepatocytes in response to antigen binding.
   c. Participates in both innate and adaptive immune responses.
   d. Prevents lysis of virus-infected cells.
   e. All of the above
4. Complement receptors
   a. Activate complement by removing peptides from inactive precursors.
   b. Allow the cell binding complement to be lysed.
   c. Neutralize viruses.
   d. Signal effector cells to phagocytose complement-coated antigen.
   e. Allow macrophages to become infected with EBV.

Short Answer

5. Name three molecules that can activate the alternative complement cascade.
6. Describe three effector functions of complement that help the body combat microbial infection.

Topic Test 1: Answers

1. **True.** Antigen-binding exposes complement-binding sites on IgM and IgG.
2. **True.** CR3 and CR4 are the adhesion molecules MAC-1 and p150.95.
3. **c.** Complement proteins are made constitutively and the alternative complement cascade can be activated immediately upon contact with bacteria, so participates in innate immunity. However, the classical complement pathway depends on antibody–antigen binding for activation, so complement also participates in adaptive immunity. Complement proteins are not active proteases in the absence of antigen.
4. **d.** Both the CR1 and the C1q receptors signal phagocytes to engulf antigen–antibody complexes. The EBV receptor, CR2, is on B cells.
5. Some of the microbial components that can activate the alternative complement cascade include lipopolysaccharide, teichoic acid, and zymosan.
6. Anaphylatoxins C3a, 4a, and 5a stimulate inflammation to attract phagocytes to the site of infection and activate their killing mechanisms. Complement opsonin C3b promotes

pathogen phagocytosis by macrophages and neutrophils, which then kill the pathogen in the phagolysosomes. The MAC is able to kill some pathogens directly by osmotic lysis. Complement coating can also block virus binding to host cells, preventing infection.

TOPIC 2: COMPLEMENT CASCADES

KEY POINTS

✓ *How does the classical complement cascade produce a C5 convertase complex?*

✓ *How does the alternative complement cascade produce a C5 convertase complex?*

✓ *During complement activation, when are opsonins and anaphylatoxins produced?*

✓ *How is MAC produced?*

The classical complement pathway is activated by antigen-antibody complexes. When antibody Fab binds antigen, Fc conformation is altered so it can bind C1q. C1q must bind at least 2 Fc regions to be activated, so activation requires two IgG molecules but only one IgM. IgA, IgE, and IgD do not bind C1q. C1q activation initiates the classical complement cascade. Antibody binding changes C1q conformation to activate C1r, which cuts a peptide bond to activate C1s. Activated C1s activates C2 and C4 by cutting a small peptide (**split fragment**) from each. An active thioester bond on C4b is exposed and covalently binds the pathogen surface. C4b associates with C2b, which has protease activity; C4b2b is called C3 convertase because it converts inactive C3 to active C3. Many molecules of C3 are cut into C3b, which acts as an opsonin, and C3a, which diffuses away and stimulates inflammation. Some C3b molecules associate with C4b2b and allow it to bind C5; C4b2b3b is called C5 convertase.

The alternative complement pathway can be activated by foreign pathogens in the absence of antibody, providing a rapid defense against certain pathogens. Small amounts of C3b are always present in body fluids due to serum and tissue protease activity. Cells with high levels of membrane sialic acid inactivate C3b if it binds, but bacteria with low external sialic acid bind C3b without inactivating it. Factor B binds C3b and serum protease; factor D cuts factor B to form active proteolytic Bb. C3bBb is the alternative pathway C3 convertase, rapidly producing more C3b and C3a. C3b joins the complex to form C3bBb3b, the alternative pathway C5 convertase. C3bBb3b is stabilized by the plasma protein properdin; the alternative pathway is also called the properdin pathway.

C5 can be cleaved by either C5 convertase into C5a and C5b. C5b combines with C6 and C7 in solution, and the C567 complex associates with pathogen lipid membrane via hydrophobic sites on C7. C8 and several molecules of C9, which also have hydrophobic sites, join MAC. Poly-C9 forms a pore in the membrane through which water and solutes can pass, resulting in osmotic lysis and cell death. If complement is activated on an antigen without a lipid membrane to which the C567 can attach, MAC may bind nearby cells and initiate bystander lysis. Gram-negative bacteria and enveloped viruses are generally more susceptible to complement-mediated lysis than are gram-positive bacteria, whose plasma membrane is protected by their thicker peptidoglycan layer. A single MAC can lyse an erythrocyte, but nucleated cells endocytose MAC and repair the damage unless multiple MACs are present.

Soluble split products C3a, C4a, and C5a are called anaphylatoxins because of their inflammatory activity. They bind receptors on various cell types to stimulate smooth muscle contraction, increase vascular permeability, and activate mast cells to release inflammatory mediators. C5a is also chemotactic for monocytes and neutrophils, and it stimulates leukocyte adherence to blood vessel walls at the site of infection, phagocytosis, and bacteriocidal functions of leukocytes.

Topic Test 2: Complement Cascades

True/False

1. Gram-positive and gram-negative bacteria are equally sensitive to complement-mediated lysis.
2. Peptides split from complement proteins during activation signal blood vessel endothelial cells to let more complement and antibodies into the infection site.

Multiple Choice

3. The classical complement cascade
   a. Is completely independent of the alternative complement cascade.
   b. Is initiated by IgM only after its Fab regions bind antigen.
   c. Must begin before the alternative pathway to provide the initial C3b.
   d. Requires at least two IgM molecules for activation.
   e. Occurs only after C1q binds Fab regions of two IgG molecules.
4. Formation of the MAC complex on the surface of the antigen
   a. Attracts leukocytes to the antigen.
   b. Is required for inflammation.
   c. Is the final step in complement-mediated opsonization.
   d. Kills pathogens by releasing their lytic granules.
   e. Results from both the classical and alternative pathways.

Short Answer

5. Why have two complement activation pathways?
6. What molecules are common to both the classical and the alternative complement pathways?

Topic Test 2: Answers

1. **False.** Both gram-positive and gram-negative bacteria have surface molecules that can activate complement, but gram-positive bacteria are less sensitive to lysis because the thick peptidoglycan cell wall blocks MAC access to the plasma membrane.
2. **True.** The split products C3a, 4a, and 5a are anaphylatoxins.
3. **b.** Antigen binding changes the conformation of mu and gamma Fc regions so they can bind and activate C1q. Classical and alternative complement cascades share some molecules, but their activation is independent.
4. **e.** Both the classical or alternative complement cascades produce C5 convertases, which activate C5b and allow it to bind the terminal complement proteins and form MAC. MAC is not involved in inflammation or opsonization.

5. The alternative complement pathway functions early in infection before antibody has been made; however, some microorganisms, particularly those with capsules, do not activate complement. Antibodies to capsular polysaccharides activate the classical pathway, which occurs later in infection but with more microorganisms.
6. Both the alternative and classical cascades use C3 and C5-9.

TOPIC 3: REGULATION OF COMPLEMENT ACTIVITY

KEY POINTS

✓ *What ensures that antibody activates complement only on the antigen surface?*

✓ *What plasma and membrane proteins regulate complement activity?*

✓ *What are the physiological effects of a deficiency in complement or its regulatory molecules?*

Complement activity is regulated by serum levels of complement components, natural decay of the activated fragments, serum protease inhibitors, and specific complement inhibitors. Serum complement levels, especially C3, often drop during infection as complement is activated faster than it is produced. The active thioester bonds on several complement fragments are short lived and inactivated by water if they do not rapidly bind the antigen. Numerous inhibitory molecules are also present in plasma and on host cell surfaces (**Table 8.2**). C1 inhibitor (C1INH) is a plasma protein that dissociates activated C1r and C1s from C1q and blocks spontaneous activation of C1 by plasma proteases. C1INH deficiency is associated with edema (swelling), described in the clinical correlation below. Several inhibitory proteins dissociate C3 and C5 convertases

Table 8.2 Complement Regulators

REGULATOR	LOCATION	FUNCTION
C1 INH	Plasma	Dissociates activated C1
C4 binding protein (C4BP)	Plasma	Dissociates classical C3 convertase Co-factor for factor I
Properdin	Plasma	Stabilizes C3bBb3b
Factor H	Plasma	Dissociates alternative C3 convertase Co-factor for factor I
Complement receptor 1 (CR1, CD35)	Membrane	Dissociates C3 convertase Co-factor for I
Decay accelerating factor (DAF, CD55)	Membrane	Dissociates C3 and C5 convertases Co-factor for factor I
Factor I	Plasma	Degrades C4b and C3b
Membrane co-factor protein (MCP, CD46)	Membrane	Co-factor for factor I
Serum proteases (anaphylatoxin inhibitor)	Plasma	Inactivate anaphylatoxins
S protein (vitronectin)	Plasma	Blocks membrane binding of soluble C567
Homologous restriction factor (HRF, C8BP, MIP)	Membrane	Inhibition of MAC
MIRL (protectin, CD59)	Membrane	Inhibition of MAC

and promote degradation of C4b and C3b by factor I, a plasma protease. These include plasma proteins C4BP and factor H and membrane proteins complement receptor 1 (CR1), decay accelerating factor (DAF), and membrane co-factor protein. Other regulatory proteins inactivate anaphylatoxins C3a, C4a, and C5a and block formation of MAC on host cells.

Deficiencies have been identified in all complement factors except 9, including factor D and properdin. Deficiencies have also been identified in complement regulatory proteins C1INH, factors I and H, DAF, and homologous restriction factor (HRF). Deficiencies in complement proteins are associated with increased bacterial infections (especially with *Neisseria*) due to reduced opsonization and phagocytosis. Immune complex disease caused by complement-mediated inflammation in response to persisting antigen–antibody complexes also increase. Regulatory protein deficiencies result in similar symptoms as complement proteins are depleted at accelerated rates. Deficiencies in factor H, DAF, and HRF also cause complement-mediated lysis of erythrocytes and appearance of hemoglobin in the urine. Complement proteins are quantified by ELISA and complement activity is measured by complement fixation (Chapter 2).

Topic Test 3: Regulation of Complement Activity

True/False

1. Deficiencies in either complement proteins or complement inhibitors can result in increased incidence of bacterial infections.
2. Complement levels rise as an infection progresses, as synthetic rates are increased in response to infection.

Multiple Choice

3. As complement is activated by complexes of antibody-coated bacteria, bystander lysis of nearby host cells is prevented by
   a. A long-lived thioester bond on active complement proteins.
   b. Covalent attachment of all active complement proteins to the pathogen surface.
   c. Plasma proteins that inactivate the anaphylatoxins.
   d. Proteins on host cell membranes that inactivate C3 and C5 convertases.
   e. The slow catalytic rates of complement proteases.
4. Complement activity is *not* inhibited by
   a. Dissociation of C3 and C5 convertases.
   b. Host cell membrane proteins that block MAC formation.
   c. Host cell plasma proteins that inactivate C3a, C4a, and C5a activity.
   d. Proteolytic cleavage of complement proteins into smaller fragments.
   e. All of the above

Short Answer

5. Why are there so many complement regulators?
6. How is complement function measured?

Topic Test 3: Answers

1. **True.** Increased incidence of bacterial infections is the most common sign of a complement deficiency.
2. **False.** Complement levels fall as an infection progresses, as complement is depleted through activation faster than it is synthesized.
3. **d.** Bystander lysis of nearby host cells is prevented by membrane inhibitors. The active split products do not bind to the pathogen, and anaphylatoxin inactivation does not affect MAC.
4. **e.** All of the above are used to inactivate complement.
5. The complement cascade, like other biological cascades, is activated very quickly and has powerful biological activities. Regulators for each part of the process ensure that complement does not damage the host cells in the absence of antigen.
6. Complement function is generally assayed by lysis of erythrocytes.

CLINICAL CORRELATION: ANGIONEUROTIC EDEMA

Angioneurotic edema, one of the most common immunodeficiency diseases, results from reduced synthesis and increased breakdown of C1INH. C1INH regulates plasma serine proteases that activate the kallikrein cascade. In C1INH deficiency, vasoactive peptides, including bradykinin and a C2a breakdown product (C2 kinin), are generated and promote capillary permeability and fluid movement into tissues. Swelling is generated by trauma, exercise, temperature extremes, or stress. Fluid in the abdomen causes abdominal pain and vomiting. Complement C2 and C4, whose activation is normally regulated by C1INH, are depleted in C1INH deficiency, but the alternative complement cascade is functional and persons with C1INH deficiency do not usually suffer from increased bacterial infections. C1INH deficiency is usually inherited but can be acquired as the result of a protracted bacterial infection, B-cell malignancy, or generation of an autoantibody to C1INH.

DEMONSTRATION PROBLEM

Complete the table below by predicting the effects of a complete deficiency of each of the complement proteins listed at the top of the chart on the activities shown in the left hand column. Use the notation N, no inhibition; P, partial inhibition; C, complete inhibition.

	C1	C3	C4	C5	C9	Factor B	C1INH
Activation of alternative C3 convertase							
Activation of classical C3 convertase							
Activation of alternative C5 convertase							
Activation of classical C5 convertase							
Complement-mediated phagocytosis							
Complement-mediated inflammation							
Complement-mediated lysis							

SOLUTION

	C1	C3	C4	C5	C9	Factor B	C1INH
Activation of alternative C3 convertase	N	C	N	N	N	C	N
Activation of classical C3 convertase	C	C	C	N	N	N	P
Activation of alternative C5 convertase	N	C	N	N	N	C	N
Activation of classical C5 convertase	C	C	C	N	N	N	P
Complement-mediated phagocytosis	P	C	P	N	N	P	P
Complement-mediated inflammation	P	C	P	P	N	P	P
Complement-mediated lysis	P	C	P	C	C	P	P

Chapter Test

True/False

1. Recurrent bacterial infections are often seen in people with low complement levels.
2. Histamine released in response to C5a binding stimulates phagocytosis of bacteria by mast cells.
3. Most of the C3b molecules produced during complement activation bind to the pathogen surface and promote binding to macrophage and neutrophil CR1.
4. MACs formed by the classical and alternative complement cascades are identical.
5. Factor I is a membrane protease that protects host cells from complement damage by dissociating MAC pores.

Multiple Choice

6. Complement is
   a. Activated by binding to specific complement receptors.
   b. Antigen specific.
   c. A potent promoter of virus entry into host cells.
   d. A series of intracellular proteins that work with antibody to eliminate endogenous antigen.
   e. Present in the circulation in an inactive form.
7. The alternative pathway of complement activation
   a. Causes tissue damage in the absence of C1INH.
   b. Occurs after the classic pathway is activated.
   c. Occurs only if the classical pathway is ineffective in pathogen clearance.
   d. Requires C3.
   e. Requires C4.
8. If a person is born without C2 and C4,
   a. C5 can still be cleaved by the classical pathway.
   b. C3b will not be able to bind to bacteria.
   c. C9 will polymerize inappropriately and lyse host cells.
   d. The classical pathway will be changed into the alternative pathway.
   e. The amount of C3b produced during bacterial infections will be reduced.

9. In the membrane attack phase of the classical complement pathway, the role of C5b is to
   a. Activate the C5 convertase activity.
   b. Attract neutrophils to lyse the pathogen.
   c. Initiate formation of the MAC.
   d. Polymerize into a membrane-spanning channel.
   e. All are activities of C5b

10. Complement receptors
   a. Activate complement on the surface of pathogens.
   b. Bind only activated complement proteins.
   c. Inhibit complement activation on the surface of host cells.
   d. On erythrocytes remove immune complexes from the circulation.
   e. On macrophages signal host cells to make opsonins.

Short Answer

11. Why is C3 used to determine someone's immune status? Does the amount go up or down during a vigorous immune response?

12. How are host cells protected from complement attack?

13. What is the primary symptom of hereditary angioneurotic edema, which affects humans with a deficiency in C1INH?

14. Both CR1 and C1qR are present on macrophages and neutrophils and signal the cells to engulf complement-coated antigen; how are they different?

Essay

15. Which complement protein is most important for complement function? Explain your answer.

Chapter Test: Answers

1. **T** 2. **F** 3. **T** 4. **T** 5. **F** 6. **e** 7. **b** 8. **e** 9. **c** 10. **d**

11. The levels of C3 are most often used to assay one's immune status because it is present in the plasma in the highest amount of any complement protein and its levels drop during a vigorous immune response to infection.

12. Host cells are protected by membrane-bound and plasma proteins that inactivate complement.

13. The primary symptom of hereditary angioneurotic edema is a nonitchy swelling (edema), often of the face and neck.

14. CR1 promotes phagocytosis only in conjunction with FcR, so it works for IgG-coated antigen. C1qR promotes phagocytosis without engagement of FcR, so it works for complement-coated and IgM-coated antigen.

15. C3 is the crucial complement protein because of its key functions in both the classical and alternative complement cascades and all three effector functions. C3b is part of the

alternative C3 convertase and both C5 convertases, necessary to activate the terminal complement pathway and produce MAC. C3b is a potent opsonin, whereas the split product C3a is an anaphylatoxin. C1, C2, and C4 participate only in the classical pathway. Factors B, D, and properdin participate only in the alternative pathway. C5 to C9 participate in complement-mediated lysis but not in opsonization; C5a is the most potent anaphylatoxin, but C3a and C4a also have anaphylatoxin activity.

Check Your Performance:

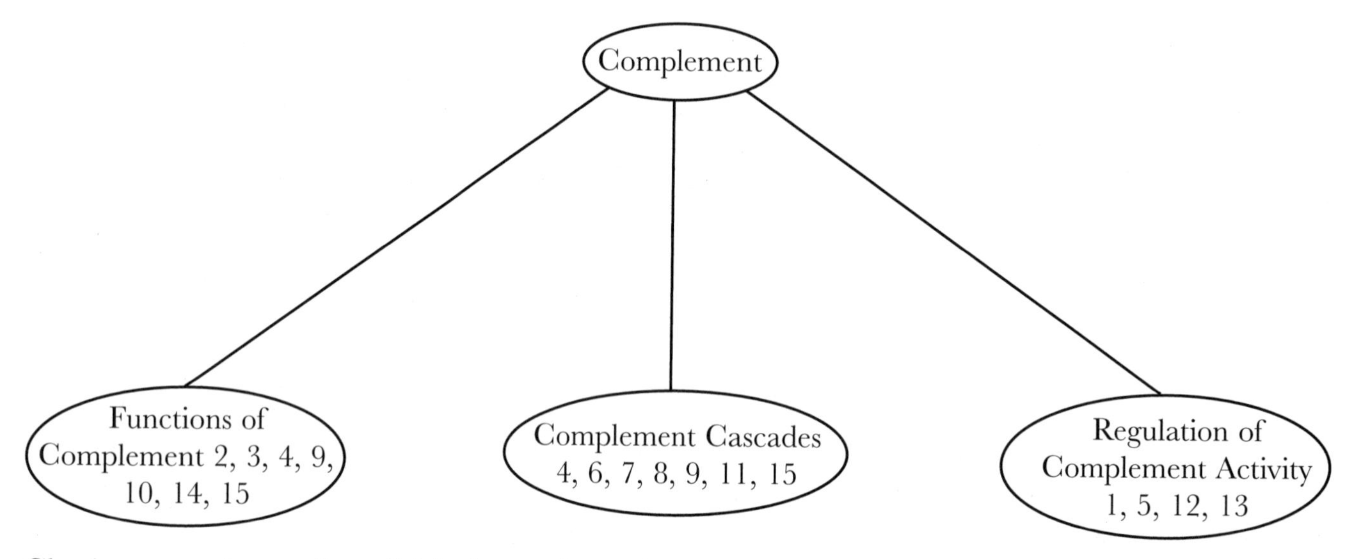

Check your understanding of this chapter by noting the number of questions in the chapter test you missed for each topic.

Innate Immunity and Inflammation

Innate immune responses are not antigen specific and do not exhibit immune memory. Phagocytosis and alternative complement activation commence immediately upon antigen contact and induce other antigen-nonspecific responses such as cytokine production, synthesis of acute phase proteins, and inflammation. Innate immune effectors often participate in initiation of acquired immune effector functions as accessory cells and molecules.

ESSENTIAL BACKGROUND

- **Immune system cells and molecules (Chapter 1)**
- **Antigen processing and presentation (Chapter 5)**
- **Chemokines and inflammatory cytokines (Chapter 7)**
- **Alternative complement activation (Chapter 8)**

TOPIC 1: ADHESION MOLECULES

KEY POINTS

✓ *What are the functions of immune system adhesion molecules?*

✓ *What structural characteristics define families of immune adhesion molecules?*

✓ *How is adhesion molecule expression regulated?*

Cell adhesion molecules (**CAMs**) are involved in cell–cell attachment throughout the body; immune CAMs direct leukocyte recirculation and promote leukocyte activation. CAMs allow mature T and B cells leaving the thymus and marrow to find the T and B cell areas in secondary lymphoid organs. On blood vessel endothelium CAMs called **vascular addressins** announce to recirculating leukocytes their location and proximity to an infection site. On lymphocytes and macrophages, CAMs also stabilize cell–cell binding so that antigen can be presented and cytokine and co-stimulatory signals can be exchanged. Some CAMs are expressed constitutively, whereas inflammatory cytokines up-regulate expression of others. Examples of major groups of immune CAMs are listed in **Table 9.1**.

Selectins are CAMs that resemble lectins in their ability to bind carbohydrate. Three kinds of selectins are responsible for leukocyte binding to vascular endothelium before their movement into the tissues: L-, E-, and P-selectins. Selectins bind carbohydrates on the mucin-like family of adhesion molecules. For example, T cells entering peripheral lymph nodes use the homing recep-

Table 9.1 Cell Adhesion Molecules of the Immune System

Name	Found On	Binds To
Selectin family		
L-selectin (CD62L, LECAM-1, MEL14)	Lymphocytes, monocytes, neutrophils, eosinophils	GlyCAM-1, CD34, MAdCAM-1 on HEV
E-selectin (PADGEM, CD62P)	Activated endothelium	Sialyl Lewis^x
P-selectin (ELAM, CD62E)	Activated endothelium	Sialyl Lewis^x PSGL-1
Mucin-like family		
CD34	HEV	L-selectin
GlyCAM-1	Endothelium	
CLA or ESL-1	Effector T cells	E- and P-selectin on activated endothelium
PSGL-1	Neutrophils	
MAdCAM-1	Mucosal endothelium	L-selectin, LPAM
Integrin family		
LFA-1 (CD11a/CD18)	T cells, macrophages, neutrophils, dendritic cells	ICAM-1, ICAM-2, ICAM-3
Mac-1 (CR3, CD11b/CD18)	Macrophages, neutrophils	ICAM-1, C3bi, fibrinogen
p150.95 (CR4, CD11c/CD18)	Macrophages	C3bi
VLA-4	Lymphocytes, monocytes, macrophages	VCAM-1
LPAM-1	Lymphocytes	MAdCAM-1
Ig superfamily		
ICAM-1	Vascular endothelial cells, lymphocytes, dendritic cells	LFA-1
ICAM-2		
ICAM-3		
LFA-2 (CD2)	T cells	LFA-3 (CD58)
LFA-3 (CD58)	APC	LFA-2 (CD2)
MAdCAM-1	Mucosal endothelial cells	Integrins and selectins on lymphocytes
VCAM-1	Vascular endothelial cells	VLA-4

tor L-selectin to recognize mucin-like vascular addressins on specialized vascular endothelium (HEV—high endothelial venules). Selectins and mucin-like CAMs are present on both leukocytes and vascular endothelium.

Integrins are leukocyte membrane CAMs that bind immunoglobulin (Ig) superfamily CAMs on vascular endothelium. Activated T cells lose L-selectin and acquire the integrin VLA-4 (very late activation antigen), which binds VCAM-1 (vascular cell adhesion molecule) on activated vascular endothelium near the site of infection. Effector T cells use a combination of CAMs to home to peripheral, mucosal, or skin endothelium. Integrin–Ig superfamily CAM interactions also stabilize interactions between leukocytes. LFA-1, CD2, and ICAM-3 (intercellular adhesion molecule) on T cells bind ICAM-1, ICAM-2, LFA-3 (leukocyte function–associated antigen), and LFA-1 on antigen-presenting cells (APC) to stabilize major histocompatibility complex (MHC)–peptide–T-cell receptor binding; similar interactions occur during cytotoxic T lymphocyte (T_C)-target binding. Integrins LFA-1, macrophage integrin–1 (MAC)-1 (CR3), and p150.95 (CR4) are dimers that share a beta chain but have distinct alpha chains. Mucosal addressin cell adhesion molecule (MAdCAM-1) has both Ig-like and mucin-like domains and binds both selectins and integrins.

Leukocytes leave the circulation and enter the tissues by a process called **extravasation.** Inflammatory cytokines signal vascular endothelium to increase CAM expression. Neutrophils usually arrive first, and their mucin-like CAMs or a sialic acid containing carbohydrate called sialyl Lewis^x binds loosely to E-selectins on the blood vessel walls. The neutrophils roll as these loose associations are repeatedly made and broken. Chemokines from the vascular endothelium and nearby inflammatory cells, as well as complement C5a and platelet-activating factor, bind receptors on the neutrophil, signaling it to activate its integrins by changing their conformation. Activated integrins LFA-1 and MAC-1 then bind more tightly to Ig superfamily ICAM-1 on the vascular endothelial cells until the neutrophils stop rolling and move between endothelial cells into the tissues. Other leukocytes extravasate in a similar manner, although some of the CAMs involved may differ.

Topic Test 1: Adhesion Molecules

True/False

1. Neutrophils arriving at an infection site bind tightly to E-selectins on vascular endothelium to move into infected tissues.
2. All leukocytes use the same CAMs to interact with vascular endothelium.

Multiple Choice

3. Lymphocyte recirculation is involved in
   a. Bringing together antigen and rare antigen-specific lymphocytes.
   b. Homing of effector T_C to a virus infection site.
   c. Homing of naive T cells to T-cell areas in the lymph nodes.
   d. Movement of mature T and B cells from the primary to secondary lymphoid organs.
   e. All of the above
4. Vascular addressins
   a. Allow lymphocytes to recognize certain tissues and return there.
   b. Are integrins.
   c. Are present on leukocyte membranes.
   d. Participate in T-cell–macrophage adhesion during antigen presentation.
   e. None of the above

Short Answer

5. How do chemokines participate in extravasation?
6. What are the four families of immune adhesion molecules and their ligands?

Topic Test 1: Answers

1. **False.** Binding to E-selectin is loose enough to slow the neutrophils to a roll; integrin–Ig superfamily CAM binding is tight and promotes extravasation.
2. **False.** Leukocytes use a combination of several CAMs to interact with vascular endothelium.
3. **e.** Lymphocyte recirculation allows naive and effector lymphocytes to contact antigen and antigen-presenting cells.

4. **a.** Vascular addressins are adhesion molecules on blood vessel endothelial cells in particular locations (like lymph node HEV) or expressed during an inflammatory response to a nearby antigen.
5. Chemokines bind their receptors on rolling leukocytes and activate integrins to bind tightly to Ig superfamily CAMs, promoting extravasation.
6. Selectins, mucin-like CAMs, integrins, and Ig superfamily CAMS: selectins bind mucin-like CAMs; integrins bind Ig superfamily CAMs.

TOPIC 2: ANTIGEN ELIMINATION

KEY POINTS

✓ *What molecules attract phagocytes to antigens?*

✓ *What molecules allow antigen to adhere to phagocytes?*

✓ *By what mechanisms are pathogens killed by phagocytes?*

Antigen removal by **phagocytosis** begins with **chemotaxis**. Phagocytes are attracted to the site of antigen or injury by chemotactic factors, including chemokines, C5a, and peptides such as fMet-Leu-Phe released from bacteria. Macrophages and neutrophils with specific receptors for the chemotactic molecules extravasate at the site of the antigen, after a concentration gradient of chemotactic molecules.

Phagocytes must first bind antigen before phagocytosis occurs. Macrophages have surface molecules that bind common microbial components. Examples include integrins MAC-1 (CR3) and p150.95 (CR4), which bind several bacterial molecules including lipopolysaccharide (LPS); mannose receptor; scavenger receptor, which binds sialic acid; and CD14, which binds LPS. Encapsulated bacteria are often resistant to phagocytosis unless they are opsonized by complement or antibody. Once antigen is bound, the phagocyte extends pseudopodia around the antigen and engulfs it, forming a phagocytic vesicle (**phagosome**). Lysosomes containing digestive enzymes at low pH fuse with the phagosome (**phagolysosome**) and their enzymes digest the antigen. Some bacteria and viruses are resistant to hydrolytic enzymes and low pH and can escape from or even live in the phagolysosome.

Phagocytes have two oxygen-dependent killing systems (**oxidative burst**) that kill microorganisms by oxidizing and inactivating key enzymes. Macrophages depend primarily on the peroxidase-independent system, using hydroxyl radicals (OH^-), superoxide anions ($^{\bullet}O_2$), singlet oxygen ($^{\bullet}O$), and hydrogen peroxide (H_2O_2). Neutrophil myeloperoxidase interacts with H_2O_2 plus intracellular halides to form toxic oxidants such as OCl^-. Activated macrophages also use nitric oxide, defensin peptides, and lysozymes to kill bacteria.

Topic Test 2: Antigen Elimination

True/False

1. Complement and complement receptors participate in phagocyte antigen binding.
2. Encapsulated bacteria are easily phagocytosed because their sticky capsule binds tightly to macrophages.

Multiple Choice

3. Phagocytosis must be preceded by
   a. Antigen binding.
   b. Chemotaxis.
   c. Extravasation.
   d. Integrin binding to Ig superfamily CAMs.
   e. Oxidative burst.

4. Pathogens engulfed by macrophages
   a. Are completely degraded by hydrolytic enzymes.
   b. Are degraded to small peptides and carbohydrates presented on class I.
   c. May survive and replicate in the macrophage phagocytic vesicles.
   d. Stimulate macrophages to adhere to B cells.
   e. Stimulate vascular endothelium to upregulate selectin expression.

Short Answer

5. How do phagocytes recognize antigen?

6. What is oxidative burst?

Topic Test 2: Answers

1. **True.** C3b, CR3, and CR4 participate in antigen binding by phagocytes.

2. **False.** Encapsulated bacteria are difficult to phagocytose because their capsule does not bind to macrophage membrane receptors.

3. **a.** Phagocytosis must be preceded by antigen binding. It *may* be preceded by chemotaxis, integrin binding to Ig superfamily CAMs, and extravasation if the phagocytes are not already present in the tissues, and it is often followed by oxidative burst.

4. **c.** Pathogens engulfed by macrophages may survive and replicate in the macrophage phagocytic vesicles. Neutrophils completely degrade antigen; macrophages present peptides but not carbohydrates on MHC class II.

5. Phagocytes recognize antigen using receptors for LPS, mannose, sialic acid, and other common bacterial surface molecules and for complement and Ig Fc.

6. Oxidative burst is a series of enzymatic reactions that produce oxygen radicals and peroxides that kill bacteria and can kill nearby host cells.

TOPIC 3: INNATE AND EARLY INDUCED RESPONSES

KEY POINTS

✓ *What effector functions are part of the innate immune response?*

✓ *What is the acute phase response?*

✓ *What are the functions of interferons and natural killer cells in early induced immunity?*

Physical and chemical barriers, including the skin and mucosal epithelia, cilia, mucus, low pH, and normal flora, usually prevent pathogens from entering the body. Pathogens that do penetrate these barriers encounter the **innate immune response**, mediated by preformed antigen-nonspecific cells and effector molecules. If pathogens are not eliminated by the innate response, an **early induced response** begins within 4 hours. The early induced response is also antigen nonspecific, but cells must be activated to synthesize effector molecules. Both the innate and early induced responses occur at the site of infection. The **adaptive immune response** involving antigen-specific lymphocytes and antibodies is not seen before 48 hours after antigen contact and is initiated in organized lymphoid tissues near the infection site.

Innate immune responses include alternative complement activation by pathogen surface molecules and phagocytosis. Complement activation leads to pathogen opsonization, lysis, and production of chemotactic and inflammatory split products. Phagocytosis results in destruction of pathogen in the phagolysosome. Macrophages are stimulated by bacterial binding to secrete inflammatory cytokines and lipid mediators of inflammation such as prostaglandins and leukotrienes, and they undergo oxidative burst as part of the early induced response. Macrophages also process and present antigen on MHC class II and express co-stimulatory molecules that allow them to activate helper T cells if an adaptive response is made. C5a, leukotriene B4, and histamine signal endothelial cells to rapidly move P-selectin from granules to the cell surface. Key inflammatory cytokines produced by macrophages include interleukin (IL)-1, which activates nearby vascular endothelium to increase expression of CAMs and promote leukocyte extravasation; the chemokine IL-8, which activates integrin on leukocytes to bind more strongly to vascular CAMs; tumor necrosis factor alpha (TNF alpha), which activates vascular endothelium to express E-selectin and, with IL-8, activates neutrophils to be more cytotoxic; and IL-12, which activates natural killer (NK) cells to kill virus-infected cells. IL-1, IL-6, and TNF alpha also cause fever, which stimulates immune responses and inhibits replication of some bacteria and viruses, and signal the liver to produce acute phase proteins. Overproduction of TNF alpha in response to systemic gram-negative bacterial infection causes **septic shock** with systemic edema, low blood pressure, and disseminated coagulation that leads to multiple organ failure and death.

Two important **acute phase proteins** are C-reactive protein (CRP) and mannan-binding lectin (MBL). CRP binds microbial LPS phosphorylcholine to opsonize the microbe and activates C1q to initiate the classical complement cascade in the absence of antibody. MBL binds bacterial mannose and opsonizes bacteria for monocytes, which lack mannose receptor. MBL also activates a serine protease cleaving C2 and C4 to initiate the lectin complement cascade. Interferons (IFN) alpha and beta are made in response to virus infections by infected host cells. IFN alpha and IFN beta are secreted and bind to membrane receptors, which activate second messengers that inhibit virus replication They also increase expression of MHC class I, TAP (transporter associated with antigen processing), and proteasome components to increase antigen presentation to T_C cells. NK cells are activated by IFN alpha, IFN beta, and IL-12 to kill virus-infected cells and by IL-12 and TNF alpha to produce high levels of IFN gamma, a strong macrophage activator.

NK cells are large granular lymphocytes that are usually $CD3^-$ and $CD16^+$. They provide an important first line of defense against virus-infected and tumor cells. NK cells have NKR-P1[5] lectin-like receptors that bind carbohydrates on self cells and signal NK cells to kill. Killing of uninfected human targets by NK cells is inhibited by Ig superfamily **killer inhibitory receptors** (KIR) that recognize MHC class I alleles; in mice, the Ly49 family inhibitory receptors are lectins that bind class I. Virus suppression of class I expression and promotion of cell protein glycosylation may make infected cells more susceptible to NK-mediated lysis. NK cells have multiple KIR, and there is some suggestion that NK cells are selected in the thymus to recognize self class I alleles. NK cells also regulate growth and differentiation of stem cells and participate in transplant rejection and autoimmunity. NK cells perform antibody-dependent cell-mediated cytotoxicity (see Chapter 10).

Topic Test 3: Innate and Early Induced Responses

True/False

1. IL-1 stimulates the hypothalamus to raise body temperature.
2. Gram-negative bacteria that get into the circulation can induce such high levels of IFN alpha secretion that septic shock results.

Multiple Choice

3. NK cells
   a. Are stimulated to kill host cells via carbohydrate-binding receptors.
   b. Kill normal host cells with high levels of membrane MHC class I.
   c. Kill virus-infected cells when the virus is acquired naturally but not by immunization.
   d. Recognize virus-infected cells by the presence of viral peptide on MHC class II.
   e. Secrete the complement MAC to lyse virus-infected cells.
4. IFN alpha and beta do *not*
   a. Activate NK cells to kill virus-infected cells.
   b. Induce macrophages to increase expression of membrane MHC.
   c. Inhibit virus replication in infected cells.
   d. Make uninfected cells resistant to virus entry.
   e. Stimulate expression of molecules required for class I binding of viral proteins.

Short Answer

5. What are the benefits of fever to the infected host?
6. What are CRP and MBL and how do they participate in early induced responses?

Topic Test 3: Answers

1. **True.** IL-1, IL-6, and TNF alpha are endogenous pyrogens.
2. **False.** Septic shock results from excessive levels of TNF alpha.
3. **a.** KIR block NK killing of host cells bearing MHC class I; there is some suggestion that virus peptide on class I may stimulate killing but how it is recognized is not known.
4. **b.** Increasing macrophage expression of MHC is a function of IFN gamma.
5. Fever stimulates immune responses and inhibits pathogen replication.
6. CRP and MBL are acute phase proteins, made by the liver in response to macrophage cytokines. They are opsonins and activate the classical complement cascade.

CLINICAL CORRELATION: CHRONIC GRANULOMATOUS DISEASE

Chronic granulomatous disease (**CGD**) is an X-linked recessive disorder characterized by a defect in NADP oxidase or in cytochrome *b*, which recycles NADP. These proteins are required for the production of superoxide radicals and peroxides and for low lysosomal pH. Without NADP oxidase, macrophages and neutrophils cannot kill ingested bacteria, especially bacteria with catalase that break down low levels of peroxides. Boys with CGD suffer from abscesses of the skin and liver, pneumonia, and lymph node infections as infected phagocytes carry bacteria throughout the body. Persistent infection stimulates T_H1 cells to make inflammatory cytokines, resulting in localized inflammation and granulomas composed of T cells surrounding the infected macrophages. Children with CGD are treated with antibiotics; in severe cases they die of septicemia in early childhood.

DEMONSTRATION PROBLEM

Integrins LFA-1, MAC-1 (CR3), and p150.95 (CR4) share a $beta_2$ chain. Describe the immune response of a knock-out mouse for the common $beta_2$ chain.

SOLUTION

Knock-out mice could not express integrins LFA-1, MAC-1, or CR4 and would be deficient in inflammation and in adaptive immunity. LFA-1 and MAC-1 are used by macrophages and neutrophils to enter infected tissues. MAC-1 and CR4 , with other molecules, also allow phagocytes to bind bacteria and signal them to kill intracellular bacteria and make inflammatory cytokines. LFA-1 strengthens interactions between T cells and APC required for helper T cell activation and cytotoxic T lymphocyte (CTL) binding virus-infected cells, so both antibody production and T-cell–mediated cytotoxicity would be impaired. A $beta_2$ chain deficiency has been identified in humans with leukocyte adhesion deficiency.

Chapter Test

True/False

1. Neutrophils are usually the earliest APC to reach the site of infection.
2. CGD results from a deficiency in macrophages and neutrophil oxidative burst.
3. Innate immune responses usually occur in the circulation where complement concentrations are highest.
4. The lectin complement cascade is triggered by acute phase proteins and requires C1, C2, and C4.
5. Chemokine signals cause a conformational change in neutrophil integrins that increases their binding to vascular endothelial cells.

Multiple Choice

6. Early induced immune responses are like adaptive immunity in that they
   a. Are antigen specific.
   b. Demonstrate immune memory.
   c. Involve macrophages and complement.
   d. Involve T and B lymphocytes.
   e. Use presynthesized proteins that can be released quickly upon cell activation.

7. Selectins
   a. Are present on both leukocytes and vascular endothelial cells.
   b. Bind Ig-like vascular addressins.
   c. Include ICAM, VCAM, and MAdCAM.
   d. Participate in positive selection of B cells in the bone marrow.
   e. Select antigen-specific macrophages to extravasate into the infection site.

8. NK cells are activated to kill virus-infected cells using receptors that bind
   a. Allogeneic MHC class I molecules.
   b. Carbohydrates on membrane proteins and lipids.
   c. Ig superfamily adhesion molecules.
   d. IFN alpha and beta.
   e. IL-12.

9. Lymphocyte recirculation
   a. Activates inflammatory cytokines to promote antigen presentation to T cells.
   b. Allows B cells to go to the site of infection to produce antibody.
   c. Circulates lymphokines efficiently throughout the body.
   d. Occurs for both naive and effector lymphocytes.
   e. Only occurs during an infection.

10. Phagocytes do *not* kill bacteria using
    a. H_2O_2.
    b. Hydrolytic enzymes.
    c. Low pH.
    d. Lysozymes.
    e. Strong reducing agents.

Short Answer

11. Against what kinds of pathogens do NK cells provide protection?
12. Describe the process of neutrophil extravasation.
13. What are the systemic effects of cytokine production by macrophages?
14. What general kinds of receptors do macrophages use to bind and engulf pathogens?

Essay

15. *Fusobacterium necrophorum*, gram-negative bacteria that live in the mouth, enter your bloodstream during routine dental work and are sequestered in the spleen, where they begin to multiply. Describe the innate and early induced immune responses to these bacteria.

Chapter Test: Answers

1. **F** 2. **T** 3. **F** 4. **T** 5. **T** 6. **c** 7. **a** 8. **b** 9. **d** 10. **e**

11. NK cells kill virus-infected cells and some enveloped viruses.

12. Neutrophil mucin-like CAMs, or sialyl Lewis^x, bind loosely to E-selectins on the blood vessel walls. Neutrophils roll as these loose associations are repeatedly made and broken. Chemokines from vascular endothelium and nearby inflammatory cells, complement C5a, and platelet-activating factor bind receptors on the neutrophil, signaling it to activate its integrins by changing their conformation. Activated integrins LFA-1 and MAC-1 bind tightly to ICAM-1 on vascular endothelial cells so neutrophils stop rolling and move between the endothelial cells into the tissues.

13. Systemic effects of macrophage cytokines include fever and liver acute phase protein synthesis. Overproduction of TNF alpha may cause systemic edema and shock.

14. Macrophage surface molecules that bind common microbial components include integrins MAC-1 (CR3) and p150.95 (CR4), which bind several bacterial molecules including LPS; mannose receptor; scavenger receptor, which binds sialic acid; and CD14, which binds LPS.

15. As soon as the bacteria enter the body, the alternative complement cascade is activated by bacterial surface molecules. Macrophages and neutrophils bind the bacterium using C3b and LPS receptors, engulf them, and kill them with low pH, hydrolytic enzymes, toxic oxygen products, nitric oxide, lysozyme, and defensin proteins. Complement MAC inserts into the bacterial outer membrane and causes osmotic lysis; split products C3a, C4a, and especially C5a stimulate mast cells to release histamine and promote localized vessel leakiness. Inflammatory cytokines produced by the macrophage promote influx of neutrophils, macrophages, and complement and cause fever and synthesis of acute phase proteins. MBL and CRP activate more complement and promote additional phagocytosis. If all the bacteria are not removed, septic shock may result.

Check Your Performance:

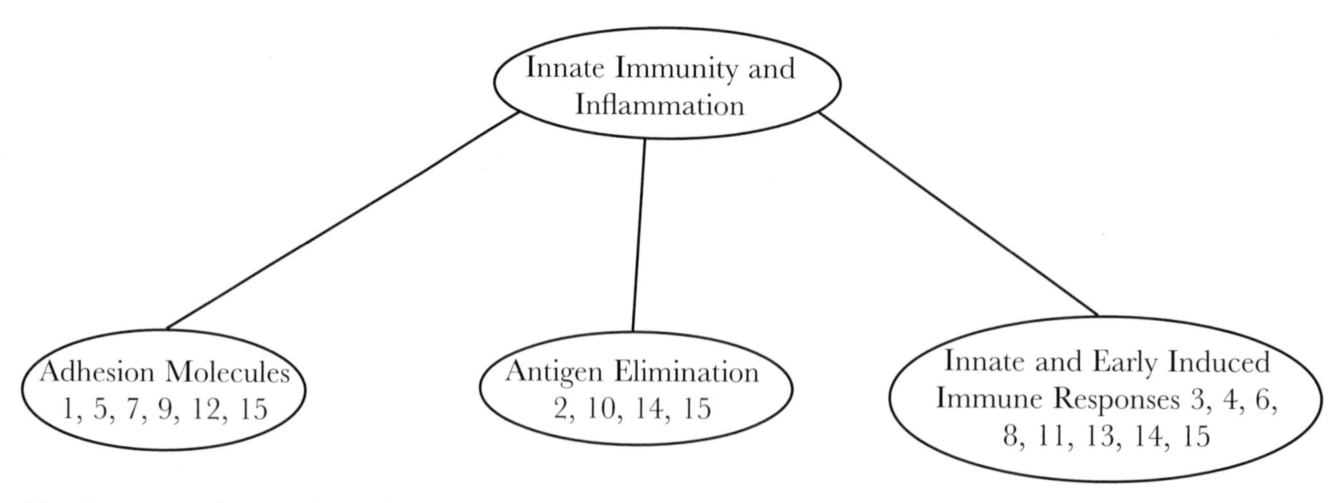

Check your understanding of this chapter by noting the number of questions you missed for each topic.

Adaptive Immunity

CHAPTER 10

Adaptive cell-mediated immunity is driven by activation of T cells: cytotoxic T cells (T_C) activated by endogenous antigen to kill infected cells and helper T cells (T_H) activated by exogenous antigen to stimulate macrophage killing of endosomal pathogens. Natural killer (NK) cells perform innate cell-mediated immunity when they kill virus-infected cells and tumor cells; they are also important for regulating hematopoiesis. Humoral immunity refers to antibody production and to the effector functions of antibody: pathogen and toxin neutralization, classical complement activation, and antibody-mediated phagocytosis.

ESSENTIAL BACKGROUND

- **Antibody structure and function (Chapter 3)**
- **Antibody genes (Chapter 4)**
- **Antigen processing and presentation (Chapter 5)**
- **T-cell receptor structure and function (Chapter 6)**
- **Cytokines (Chapter 7)**
- **Complement (Chapter 8)**
- **Innate immunity (Chapter 9)**

TOPIC 1: T-DEPENDENT AND T-INDEPENDENT ANTIGENS

KEY POINTS

✓ *What are the structural differences between T-dependent and T-independent antigens?*

✓ *How does humoral immunity to T-dependent and T-independent antigens differ?*

✓ *What are the advantages of a T-independent response to pathogens?*

Most antigens are **T dependent**; T-cell help is required for maximal antibody production. T-dependent antigens are protein because only peptides can be presented on major histocompatibility complex (MHC) class II to T_H cells, which then provide co-stimulation to trigger B-cell proliferation and differentiation into plasma cells. Class switching to IgG, IgA, and IgE and memory cell generation occur in response to T-dependent antigens.

Some antigens that cannot be presented to T cells still activate B cells to divide and secrete antibody; these antigens are **T independent** and generate only IgM antibodies. T-independent antigens give B cells both antigen and co-stimulatory signals; T-dependent antigens only deliver antigen signal to the B-cell receptor (BCR). T-independent antigens include polyclonal B cell activators (mitogens), which are often microbial surface molecules like lipopolysaccharides (LPS) or carbohydrates that bind B-cell membrane receptors and provide co-stimulatory signals. At high concentrations, these T-independent antigens stimulate B cells of many specificities to divide and secrete antibody (**polyclonal activation**). At lower concentration, only B cells whose BCR specifically bind them are stimulated to divide and secrete specific IgM. Responses to polyclonal B-cell activators can be observed in the absence of T cells and occur more quickly than T-dependent responses because no T-cell activation is required.

Other T-independent antigens have multiple repeating epitopes, for example, bacterial polysaccharide capsules. These extensively cross-link BCR on specific B cells, stimulating them to produce IgM. Antibody responses to polysaccharides are not made in infants because their less mature B cells are inactivated by BCR cross-linking. Responses to polysaccharides occur in mice lacking a thymus but not in the total absence of T cells. T cells that develop outside the thymus may be activated by these antigens, possibly presented on nonclassical MHC-like molecules, to provide some help to responding B cells.

Topic Test 1: T-Dependent and T-Independent Antigens

True/False

1. High doses of some T-independent antigens induce production of antibody to which they do not bind.

2. T-independent responses are faster than T-dependent responses.

Multiple Choice

3. T-independent antigens do *not*
   a. Bind BCR.
   b. Get presented on MHC class II.
   c. Have repeating epitopes.
   d. Induce B-cell proliferation.
   e. Provide co-stimulatory signals to B cells.

4. T-independent antigens are often
   a. Components of self-cell membranes.
   b. Polyclonal T-cell activators.
   c. Repeating peptide epitopes.
   d. Too small to be phagocytosed and presented.
   e. None of the above

Short Answer

5. What is a polyclonal B-cell activator?

6. Why don't infants make good humoral responses to bacterial polysaccharides?

Topic Test 1: Answers

1. **True.** Some T-independent antigens are polyclonal activators (mitogens).
2. **True.** Because specific B and T cells do not need to find each other, B cells can begin secreting antibody sooner.
3. **b.** T-independent antigens are not presented on MHC class II.
4. **e.** T-independent antigens are often components of bacterial surfaces. They may be polyclonal B-cell activators and have repeating nonpeptide epitopes (lipid or carbohydrate). They are often very large molecules.
5. A polyclonal B-cell activator binds to B-cell surface molecules in addition to (or instead of) BCR, signaling the B cells to divide and make IgM independent of their antigenic specificity. Examples include bacterial LPS and lectins (carbohydrate-binding molecules).
6. Infant's B cells are less mature than those of adults, and the immature B cells are usually made unresponsive when their BCR is extensively cross-linked.

TOPIC 2: ACTIVATION OF T LYMPHOCYTES

Key Points

✓ *What signals are required to activate a T cell?*

✓ *How are antigen-presenting cells stimulated to express co-stimulatory molecules?*

✓ *What roles do adhesion molecules play in T-cell activation?*

Antigen contact with mature naive T cells, plus the proper co-stimulatory signals, induces them to divide and differentiate into armed effector T cells: T_H1 cells, which activate macrophages to kill vesicular pathogens; T_H2 cells, which activate B cells to make antibody; and T_C cells, which kill cells infected with endogenous pathogens. Naive T cells encounter antigen on antigen-presenting cells (APC) in the secondary lymphoid organs. T cells use adhesion molecules to attach loosely to APC and "sample" the peptides on their MHC. Compatible peptide–MHC binding to T-cell receptor (TCR) changes the conformation of leukocyte function–associated antigen (LFA-1) to stabilize T cell–APC binding to enable activation signals to be delivered.

To be activated, naive T cells binding antigen peptide on MHC must also receive a co-stimulatory signal. **Co-stimulation** usually comes from membrane B7 on APC binding membrane CD28 on T cells; interleukin (IL)-1 from activated macrophages can also provide co-stimulation. Dendritic cells in lymphoid organs constitutively express MHC class I, class II, and B7 and can be infected by many viruses, so they are efficient APC for both T_H and T_C cells. Macrophages and B cells have low constitutive levels of B7, but contact with microbial (mostly bacterial) surface molecules signals them to increase B7 expression. The same cell generally delivers antigen plus co-stimulatory signal; if T cells do not receive co-stimulation, they become **anergic** (unable to function). This ensures that self-epitopes will usually not be able to activate self-specific T cells that escaped negative selection. Naive T_C cells receive co-stimulation from virus-infected dendritic cells or, if virus-infected cells lack co-stimulatory activity, $CD4^+$ cells induce B7 expression or produce IL-2 close to where the T_C cell binds antigen. Activated T cells express CTLA-4 instead of CD28; CTLA-4 binding to B7 inhibits T-cell function.

Binding signals at the T-cell membrane are communicated to the nucleus by way of chemical **second messengers**. Antigen binding results in TCR and co-receptor (CD4 or CD8) clustering on the T-cell membrane. Aggregation promotes association of immunoreceptor tyrosine-based activation motif (ITAM) sequences on the CD3 cytoplasmic domain with the tyrosine kinase Fyn and brings them close to the tyrosine kinase Lck associated with the co-receptor cytoplasmic domain. Fyn and Lck phosphorylate CD3 zeta and epsilon chains, which then bind another tyrosine kinase, ZAP-70. Fyn and Lck phosphorylate ZAP-70, which phosphorylates Ras to initiate a sequence of kinases that eventually activate a DNA transcription factor, Ap-1. Fyn and ZAP-70 also activate phospholipase C to cleave the membrane phospholipid phosphatidylinositol to diacylglycerol and inositol triphosphate. Diacylglycerol production activates protein kinase C, which activates transcription factor NFκB. Inositol triphosphate production results in increased intracellular Ca^{2+} and activation of calcineurin, a phosphatase that activates another transcription factor, NFAT. Ap-1, NFκB, and NFAT promote transcription of specific genes that result in T-cell proliferation and differentiation.

Activated T cells synthesize and secrete IL-2 and increase expression of high affinity IL-2R. Antigen, co-stimulatory, and IL-2 signals cause T cells to divide and differentiate into a clone of **armed effector cells** that do not need further co-stimulation to provide help or cytotoxicity. Activated T cells express cell adhesion molecules that direct them to the site of infection. An activated $CD4^+$ T cell may become an effector T_H1 or a T_H2 helper cell, depending on its cytokine microenvironment. In general, T_H1 cells stimulate cellular immunity and T_H2 cells stimulate humoral immunity. Activated $CD8^+$ cells become effector T_C lymphocytes (CTL). Memory T_H and T_C cells are also produced; they are longer lived and have reduced requirements for co-stimulation than naive T cells.

Topic Test 2: Activation of T Lymphocytes

True/False

1. Antigen-binding T cells first divide in response to IL-2 and then differentiate into effector helper or cytotoxic cells.
2. Liver cells have high levels of B7, so after their infection with hepatitis B virus they rapidly activate T_C.

Multiple Choice

3. Membrane events required for T_H activation do *not* include binding of T cell
   a. CD4 to antigen-presenting cell MHC class II.
   b. CTLA-4 to antigen-presenting cell B7.
   c. High affinity IL-2R to IL-2.
   d. LFA-1 to antigen-presenting cell ICAM (intercellular adhesion molecule).
   e. TCR to peptide on antigen-presenting cell MHC class II.
4. Cytoplasmic signals generated in response to antigen binding and co-stimulatory signals to T_C cells include all of the following *except*
   a. Activation of transcription factors that increase synthesis of IL-2.
   b. Increased free intracellular Ca^{2+}.
   c. Induction of RAG synthesis.

d. Production of second messengers from membrane inositol phosphate.
e. Tyrosine kinase binding to the cytoplasmic domains of CD3 and CD8.

Short Answer

5. Where do T helper cells first encounter antigen?
6. What is an ITAM?

Topic Test 2: Answers

1. **True.** Antigen-binding T cells first expand into clones of specific cells, which then differentiate into effector helper or cytotoxic cells.
2. **False.** Professional APC express or can be induced to express B7, but most other cells do not. The IL-2 signal for T_C proliferation often must come from activated T_H cells.
3. **b.** CTLA-4 binding to APC B7 occurs when activated T cells encounter B7, which inhibits their activity. Naive T cells are activated by CD28 binding to APC B7.
4. **c.** RAG is produced only during T-cell development, when recombination of TCR gene segments occurs.
5. T_H cells first encounter antigen on APC in the secondary lymphoid organs.
6. An ITAM is an amino acid sequence in the cytoplasmic domain of lymphocyte membrane molecules that binds cytoplasmic tyrosine kinases to promote lymphocyte activation.

TOPIC 3: HUMORAL IMMUNITY

KEY POINTS

✓ *What molecular signaling events follow antigen binding by B cells?*

✓ *How does the architecture of the secondary lymphoid organs facilitate B-cell activation?*

✓ *What processes maintain antibody specificity and allow for affinity maturation during an immune response?*

✓ *How are plasma cells and memory cells different from naive B cells?*

✓ *How do primary and secondary responses to the same antigen differ?*

✓ *How does antibody interact with innate immune effector mechanisms?*

B-cell activation is initiated by BCR cross-linking. Haptens, which are single epitopes, cannot activate B cells unless they are present on carrier molecules. Aggregation of BCR–Ig alpha/Ig beta complexes promotes ITAM phosphorylation by cytoplasmic tyrosine kinases Fyn, Blk, and Lyn to initiate second messenger cascades similar to those in T cells. Bound antigen is internalized with its BCR and processed, combined with MHC class II, and reexpressed on B-cell membranes to be presented to T_H cells. T_H cells are activated by antigen and co-stimulatory signals from professional APC, usually interdigitating dendritic cells, in the T-cell areas of lymphoid tissue. Antigen-specific B cells entering the lymphoid organs are trapped in the T-cell zone and bind the specific activated T_H cells.

B cells and T_H cells must recognize epitopes on the same antigen, but not necessarily the same epitopes, to interact. For example, a B cell that binds one virus epitope endocytoses the whole virion and presents many virus epitopes to different T_H cells, any one of which could activate the B cell. Polysaccharide T-independent antigens can be made T-dependent by linking the polysaccharides to protein; B cells specific for the polysaccharide then present the protein epitopes to T_H cells. The requirement for T- and B-cell epitopes to be physically linked on the same antigen or pathogen ensures that the immune response remains specific. Adhesion molecules hold T_H and B cells together long enough for peptide–MHC–TCR interactions, which trigger the T_H cell to co-stimulate the B cell. Binding of antigen-specific T_H2 cells and B cells increases T-cell membrane CD40 ligand (CD40L) expression and IL-4 secretion, which stimulate B-cell proliferation. T-cell cytoskeleton and Golgi move so that signals are directed toward the specific B cell.

During a primary immune response, B cells are activated by antigen plus cytokines in the paracortical areas of secondary lymphoid tissues. Activated B cells become short-lived IgM-secreting plasma cells or migrate with their specific T_H cells into **follicles** to divide and form **germinal centers** within a few days of antigen stimulation. In a primary immune response, IgM is secreted before isotype switching occurs and is the predominant isotype produced. IL-2, IL-4, and IL-5 promote B-cell division; other cytokines drive Ig isotype switching. IL-2 stimulates J chain production; interferon (IFN) gamma stimulates switching to IgG_{2a}; IL-4 stimulates switching to IgE or IgG_1; and IL-5 stimulates switching to IgA. IL-6 promotes antibody secretion.

Class switching, affinity maturation, and generation of memory B cells occur late in a primary immune response. **Class switching** is a recombinatorial event that increases the functional diversity of Ig molecules but does not affect their antigen-binding specificities. Antigen-binding specificity is altered by **somatic hypermutation**, which occurs in rapidly dividing B cells in germinal centers. B cells that remain specific for the stimulating antigen are selected by binding antigen held on Fc receptor (FcR) and complement receptors on **follicular dendritic cells** (**FDC**), which do not express MHC class II. Only B cells that bind antigen with sufficiently high affinity continue to secrete antibody and differentiate into plasma cells; low binding affinity B cells undergo apoptosis. The B-cell co-receptor complex of CR2, CD19, and TAPA 1 responds to complement or FDC CD23 binding by initiating a phosphorylation cascade that synergizes with BCR signaling, increasing the sensitivity of B cells to antigen, and with IL-1 promotes B-cell differentiation into plasma cells. **Affinity maturation** occurs by clonal selection as antigen levels drop due to antigen clearance, resulting in survival of B cells with the highest antigen-binding affinity. Further T_H cell interaction with these B cells signals them to become plasma cells or memory cells.

Plasma cells are antibody-producing cells that no longer divide or respond to antigen. They are larger than B cells and have more ribosomes, endoplasmic reticulum, and Golgi but no membrane Ig. Memory B cells are functionally and physically distinguishable from naive B cells. They often have membrane IgG, IgA, or IgE and higher levels of ICAM-1 and CR than naive B cells and are thought to live longer. Low levels of antigen may remain on FDC for years, so that B cells may be continually activated at low levels to replenish memory cell populations.

The kinetics of primary and secondary immune responses differ. The **primary** antibody response to a protein antigen is largely IgM, becoming detectable in serum 5 to 7 days after injection of antigen and dropping off after about 2 weeks. IgG may be produced late in the primary response. IgM is lower affinity but higher avidity than IgG and more efficient at complement fixation. A **secondary** (**anamnestic**, **memory**) immune response occurs after reexposure to the same antigen; the two responses may be merged when the antigen is a replicating pathogen that persists in the body. More antibody is produced during a secondary response, it is produced faster (peak at 7 days instead of 12 to 14), and relatively more IgG is produced. Early in a secondary immune response, antibody from a previous response may be present and bind antigen. **Immune complexes** (antigen + antibody ± complement) enter secondary lymphoid tissues containing memory T and B cells. Immune complexes bind FDC to activate B cells more quickly. T cells near germinal centers are activated by and provide help to nearby antigen-presenting B cells. Because of their increased frequency and antigen-presenting specificity, B cells are very efficient APC in secondary responses.

Once antibody has been synthesized, it neutralizes viruses and toxins and promotes extracellular antigen elimination. IgG on the surface of pathogens activates complement to promote inflammation and pathogen lysis. Antibody- and complement-coated antigen, especially bacterial capsules that are resistant to phagocytosis, are more readily bound by phagocytes and destroyed. **Antibody-dependent cell-mediated cytotoxicity** (**ADCC**) is performed by killer (K) cells with membrane FcR. K cells include NK cells, neutrophils, eosinophils, and macrophages. Antigen-antibody binding stimulates the K cell to release cytotoxic molecules that kill the target cell. Antigen specificity comes from the bound antibody. The cytotoxic mechanism depends on the particular K cell involved: NK cells and eosinophils use perforins and granzymes, whereas macrophages and neutrophils release proteolytic enzymes and oxidants. NK cells and activated macrophages may also release tumor necrosis factor (TNF). MHC is not involved in ADCC.

Topic Test 3: Humoral Immunity

True/False

1. Plasma cells secrete antibody as long as antigen is available to bind their membrane Ig receptors.
2. To be a K cell and perform ADCC, a cell must be have complement receptors.

Multiple Choice

3. Antigen-binding B cells entering the secondary lymphoid organs initially go to the
   a. B-cell areas where they can bind antigen presented by follicular dendritic cells.
   b. B-cell areas where they can process and present antigen to T cells.
   c. Plasma cell areas where they can secrete antibody.
   d. T-cell areas where they can be the predominant activator of naive T cells.
   e. T-cell areas where they can find specific helper T cells.
4. Affinity maturation of the humoral immune response is due to
   a. Continued stimulation of B cells by high levels of antigen on the FDC.
   b. DNA recombination by products of the RAG genes.
   c. Isotype switching.
   d. Negative selection of T cells with the lowest helper potential.
   e. Positive selection of B cells with the highest affinity for antigen.

Short Answer

5. How are B- and T-cell recognition of antigen linked?
6. How and why are primary responses to protein antigens and pathogens different?

Topic Test 3: Answers

1. **False.** Plasma cells secrete antibody as long as they live, several days to several weeks; they have no membrane Ig with which to bind antigen.
2. **False.** K cells have FcR on their membrane.
3. **e.** Antigen-binding B cells entering the secondary lymphoid organs initially go to the T-cell areas where they can find specific helper T cells, which have already been activated (usually by interdigitating dendritic cells).
4. **e.** Affinity maturation of the humoral immune response is due to positive selection of B cells with the highest affinity for antigen, especially at lower levels of antigen.
5. B- and T-cell antigen recognition are linked by the requirement for the B cell to present antigen on MHC class II to a specific T cell before the T cell can provide helper cytokines to activate the B cell.
6. Protein antigens induce distinct primary and secondary humoral responses. The primary response is predominantly IgM, which peaks at about 10 to 14 days and drops quickly as antigen is eliminated; the secondary response is predominantly IgG, which peaks at about 7 days and remains elevated for weeks. Both IgM and IgG may be made in a primary response to a pathogen and antibody titers may remain elevated longer. Pathogens replicate in the body and persist longer because they have strategies for avoiding the immune system. Protein antigens cannot increase in amount and are often cleared more quickly than pathogens, so less isotype switching may occur upon first exposure.

TOPIC 4: T-CELL–MEDIATED IMMUNITY

KEY POINTS

✓ *How do cytotoxic T effectors recognize and kill their target cells?*

✓ *How are CTL similar to and different from NK cells?*

✓ *How do T_H1 cells regulate cellular immunity?*

✓ *How does macrophage activation increase intracellular killing?*

CTL mediate antigen-specific MHC-restricted cytotoxicity and are important for killing cytoplasmic parasites inaccessible to secreted antibody and phagocytes. Examples include all viruses, rickettsiae (causes of Rocky Mountain spotted fever and typhus), and some obligate intracellular bacteria like *Chlamydia*. Effector CTL bind their targets first via adhesion molecules such as LFA-1 and then with TCR. TCR–peptide–class I–CD8 binding reorients the CTL cytoskeleton to focus release of effector molecules toward the target cell. CTL induce apoptosis in their targets. Programming the targets to die requires approximately 5 minutes of contact between CTL and target, although targets appear viable for much longer. CTL dissociate to bind and kill other target cells presenting the same epitope.

During their maturation into effector CTL, T_C cells synthesize **perforin** and **granzymes** and store them in cytoplasmic granules. Once the target cell is bound, CTL release perforin and granzymes into the space between CTL and the target cell. This directional release prevents killing of nearby uninfected cells. Perforin, similar in sequence to complement C9, also polymerizes to form pores in the target cell membrane through which water and salts enter. At high perforin concentrations, osmotic lysis may destroy the target cell; physiological levels of perforin promote apoptosis by allowing granzymes to enter the target cell. Apoptotic enzymes destroy cytoplasmic pathogens so that they cannot infect nearby cells. Dead target cells are rapidly ingested by macrophages without activating macrophage expression of B7, preventing co-stimulation of self-specific T cells.

Binding of CTL membrane Fas ligand (FasL) to target cell Fas (APO-1, a member of the TNF receptor family) induces target cell apoptosis in the absence of perforin and granzymes. Because activated T_H1 and T_H2 cells express FasL, they may also be cytotoxic. Mice with defects in Fas or FasL have a higher incidence of lymphoproliferative disease and autoimmunity, suggesting that Fas-mediated cytotoxicity may regulate immune responses. CTL also regulate immunity by releasing IFN gamma, TNF alpha, and TNF beta. IFN gamma inhibits viral replication, activates IL-1 and transporter associated with antigen processing (TAP) expression on infected cells to promote antigen processing, and recruits and activates macrophages as APC and effector cells. TNF alpha and TNF beta act with IFN gamma to activate macrophages and to directly kill some target cells.

NK cells mediate early innate cytotoxic responses to viruses and tumor cells by mechanisms similar to those used by CTL, although recognition of targets is quite different. Natural killing is antigen nonspecific and not MHC restricted; self-MHC protects targets from being lysed. NK cells do not require activation to be cytotoxic because they synthesize granzymes and perforin constitutively. However, NK cytotoxicity increases after activation by IFN alpha, IFN beta, and IL-12. NK cells use perforin and granzymes to induce apoptosis and kill multiple targets.

Some pathogens are phagocytosed but prevent fusion of lysosomes and phagosomes or acidification of the phagolysosome that activates lysosomal enzymes. Some pathogens escape the phagolysosome and live in the macrophage cytoplasm. In all cases, the pathogens are protected from complement and antibodies; peptides from the vesicular pathogens are not presented on class I MHC. Macrophage activation by armed effector T_H cells is required to eliminate these pathogens. Macrophage activation requires T-cell binding to antigen peptide on macrophage class II, sensitization of the macrophage, and activation of the macrophage by IFN gamma. T_H1 membrane CD40L binds macrophage CD40 and signals the macrophage to express receptors for IFN gamma, which is synthesized de novo by the T_H1 cell. CD40L and IFN gamma secretion are directed toward the macrophage-presenting antigen to the T_H1 cell, so that activation is generally limited to infected macrophages. Macrophages are also sensitized by LPS and activated by membrane-bound TNF alpha or TNF beta. IFN gamma produced by activated T_C cells can activate macrophages presenting cytosolic peptides on MHC class I.

IFN gamma activates macrophages to upregulate MHC class II, B7, CD40, and TNF receptor expression to recruit more effector T_H1 cells and to become more sensitive to CD40L and TNF alpha. Activated macrophages fuse their phagosomes and lysosomes more efficiently and increase their synthesis of nitric oxide, oxygen radicals, antimicrobial peptides, and IL-12. When pathogens are present that cannot be phagocytosed, activated macrophages can release oxygen radicals, nitric oxide, and proteases; these compounds kill the pathogen but also damage surrounding host tissues. Chronically activated macrophages may become less cytotoxic; if these macrophages express Fas, they can be killed by FasL-expressing T_H1 cells. Granulomas, macrophages surrounded by activated T cells, form when intracellular pathogens cannot be eliminated. The local inflammatory response resulting from activated T_H1 cells and macrophages is called delayed type hypersensitivity.

In addition to activating macrophages, T_H1 cells secrete cytokines that regulate cellular immunity. IL-2 stimulates clonal proliferation of T cells; IL-3 and granulocyte-macrophage colony-stimulating factor stimulate macrophage differentiation in the bone marrow. Multiple cytokines, including TNF alpha, TNF beta, and macrophage chemotactic protein, stimulate macrophage recruitment to the site of infection.

Topic Test 4: T-Cell–Mediated Immunity

True/False

1. Only virus-infected cells are susceptible to CTL-induced apoptosis.
2. Granulomas form when activated macrophages are not able to completely eliminate vesicular pathogens.

Multiple Choice

3. CTL binding and destruction of target cells depends on
   a. A co-stimulatory signal from the target cell.
   b. Antigen presentation on the surface of the target cell.
   c. Lack of a negative signal from MHC to prevent cytotoxicity.
   d. Presence of virus proteins in the membrane of the target cell.
   e. Secretion of cytokines by the infected cell to attract CTL.
4. Macrophage activation by T_H1 cells is an important immune mechanism for eliminating
   a. Bacteria that can resist lysosomal degradation.
   b. Bacteria whose capsule makes them resistant to phagocytosis.
   c. Enveloped viruses.
   d. Parasites that infect T cells.
   e. Viruses that infect macrophages.

Short Answer

5. When CTL lyse virus-infected cells with perforin and granzymes, what protects nearby uninfected cells from being lysed?
6. What is the role of IFN gamma released by CTL?

Topic Test 4: Answers

1. **False.** Cells infected with other cytoplasmic pathogens (like rickettsia and some bacteria) and cells bearing Fas can also be killed by CTL.
2. **True.** Granulomas are collections of T_H1 cells and macrophages in which some pathogens still exist; an example is the tubercles in tuberculosis.
3. **b.** CTL binding and destruction of target cells is initiated by binding to specific processed antigen presented on target cell MHC class I. Co-stimulation of effector CTL is not required, nor is cytokine secretion by the target.

4. **a.** Macrophage activation by T_H1 cells is an important immune mechanism for eliminating bacteria resistant to lysosomal degradation. Viruses and other cytoplasmic pathogens whose antigens are presented on class I are eliminated by CTL.
5. CTL binding targets rearrange their cytoskeleton to preferentially secrete granzymes and perforin into the space between themselves and the target.
6. IFN gamma inhibits viral replication, activates IL-1 and TAP expression on infected cells to promote antigen presentation, and recruits and activates macrophages.

CLINICAL CORRELATION: HYPER IgM SYNDROME

Hyper IgM syndrome is an X-linked immunodeficiency in which the gene for CD40L is defective. Boys with this immunodeficiency have T cells that cannot activate B cells or macrophages. They make IgM antibody to T-independent antigens like the carbohydrate blood group antigens A and B, but they cannot make IgG or IgA antibodies to protein antigens or effectively fight intravesicular pathogens like mycobacteria. They are susceptible to infections by staphylococci and streptococci, which cannot be phagocytosed efficiently without IgG opsonization, and get infected by pathogens such as *Mycobacterium kansasii* that do not normally cause disease in healthy people.

DEMONSTRATION PROBLEM

Haptens helped immunologists understand antigen recognition and T cell and B cell cooperation required for IgG synthesis. In the experiment shown below, mice were injected with the small nonprotein hapten phosphorylcholine, either unlinked (PC) or chemically linked to a protein carrier, hen egg lysozyme (PC-HEL) or human serum albumin (PC-HSA). IgG antibody to PC was measured after the second injection. Explain how the data shown below supports the requirement for T_H and B cells to recognize epitopes on the same T-dependent antigen.

Number	Primary antigen	Secondary antigen	IgG anti-PC
1	None	PC or PC-HEL	–
2	PC	PC or PC-HEL	–
3	PC-HEL	PC-HEL	+++
4	PC-HEL	PC + HEL (unlinked)	–
5	PC-HEL	PC-HSA	–
6	PC-HEL	PC-HSA + HEL	–
7	PC-HEL + HSA	PC-HSA	+++

HSA, human serum albumin; HEL, hen egg lysozyme; PC, phosphorylcholine (a hapten).

SOLUTION

IgG is generated only in T-dependent secondary responses to protein antigens. Because PC is not protein, it cannot be presented on classical MHC to bind TCR, which must recognize epitopes on the protein carrier. Therefore, PC contains the B-cell epitope and HEL or HSA contains the T-cell epitopes. No PC-specific IgG is made after a single injection of PC or PC-HEL (number 1) or a primary injection of PC followed by the hapten-carrier conjugate (number 2). The secondary antigen must contain both the T- and B-cell epitopes that generated memory cells in the primary response (numbers 3 vs. 5). Note that the hapten and carrier must be physically attached to each other in the secondary injection (numbers 3 vs. 4 and 6) to generate IgG to the hapten. However, they do not need to be physically attached in the primary injection (number 7) as long as both hapten and carrier are given in a form that will induce memory cells. [In all experiments, the B cells also made antibody to the carrier(s).]

Chapter Test

True/False

1. T-independent antigens can be made T-dependent by linking them to protein carriers.
2. Antigen-binding signals to T-cell membrane TCR are communicated to the cytoplasm by kinases present in the cytoplasmic tail sequence of TCR.
3. Because Epstein-Barr virus infects B cells, which are professional APC, it can be presented to $CD8^+$ T cells on MHC class II.
4. Ig isotype switching is a DNA recombination event.
5. Cytokines produced by T_H1 cells signal the bone marrow to produce more macrophages.

Multiple Choice

6. The earliest response to a primary infection with Epstein-Barr virus would be
   a. Antibody production.
   b. Antigen-specific T-cell cytotoxicity of virus-infected cells.
   c. Natural killing of virus-infected cells.
   d. Phagocytosis of virus-infected cells.
   e. Increased expression of co-stimulatory molecules on dendritic cells.
7. Humoral immunity exhibits a lag phase because
   a. B cells must mature to plasma cells before secreting antibody.
   b. IgG is produced immediately but in titers too small to be detected.
   c. Plasma cells must find antigen before secreting antibody.
   d. Specific T_H cells, B cells, and antigen must co-localize in the secondary lymphoid organs.
   e. T-independent antigens stimulate B cells more slowly than T-dependent antigens.
8. Plasma cells
   a. Are very long lived.
   b. Divide and differentiate into memory B cells.
   c. Produce most of their antibody at the site of infection.
   d. Secrete antibodies as long as antigen binds their membrane Ig receptors.
   e. None of the above
9. Cellular immunity includes all of the following *except*
   a. Activated macrophages killing phagocytosed bacteria.
   b. $CD8^+$ T cells inducing apoptosis using Fas-FasL binding.
   c. Complement-mediated lysis of bacterial cells.

d. MHC-restricted cytotoxic T cells killing virus-infected cells.
e. NK cells performing natural killing of tumor cells.

10. Which of the following would you expect *not* to reduce CTL killing activity?
    a. Absence of the gene for beta$_2$-microglobulin.
    b. Absence of the gene for perforin.
    c. Presence of anti-CD8 antibodies.
    d. Presence of KIR.
    e. Presence of monoclonal antibodies specific for multiple TCR Vα and Vβ specificities.

Short Answer

11. After antigen-binding B cells are stimulated by helper T cells in the lymph node paracortex, why do the B cells need to bind antigen again in the lymphoid follicles?
12. How do antigen–antibody complexes binding to FcR provide important humoral effector functions?
13. Virus-infected cells killed by CTL and phagocytosed by macrophages do not induce B7 expression on the macrophages; why is this important?
14. You discover a new bacterial pathogen that causes serious respiratory disease and death in children under 18 months old. The pathogen is most virulent when it makes a polysaccharide capsule, allowing it to resist phagocytosis. Adults are not susceptible to this pathogen because they make a T-independent IgM response. IgM fixes complement that allows the bacterium to bind complement receptor on macrophages and to be phagocytosed. Design a vaccine that can be given to infants with their first diphtheria-pertussis-tetanus (DPT) vaccination at 2 months of age.

Essay

15. Compare and contrast the cellular immune effector functions offered by CTL, macrophages, and NK cells against infectious disease.

Chapter Test: Answers

1. **T** 2. **F** 3. **F** 4. **T** 5. **T** 6. **c** 7. **d** 8. **e** 9. **c** 10. **d**

11. Rapidly dividing B cells can undergo somatic mutation of their BCR, changing its antigen specificity. A second antigen binding event allows for the positive selection of high-affinity B cells to mature and differentiate into plasma cells and memory cells, maintaining specificity of the immune response.
12. Antigen–antibody complexes binding to FcR promote antigen uptake by macrophages and neutrophils and allow K cells to bind antigen and perform ADCC.
13. Virus-infected cells are self. Without B7 expression, the macrophages cannot deliver co-stimulatory signals to activate any self-specific T_H cells that might have escaped negative selection in the thymus and bound to self on macrophage MHC class II. This reduces the risk of inducing autoimmunity.

14. Infants' B cells are too immature to make T-independent responses; they undergo clonal anergy if their BCR are extensively cross-linked. The most effective vaccine would be one with the polysaccharide epitope covalently linked to a protein carrier to induce a T-dependent IgG response that could also activate complement and promote phagocytosis. Because DPT will be given anyway, the polysaccharide could be linked to diphtheria toxoid or tetanus toxoid, which are protein antigens; the complex would then induce a memory response to both the diphtheria toxoid (DT) or tetanus toxoid (TT) and the polysaccharide antigen.

15. CTL, macrophages, and NK cells all use cytotoxicity to protect against intracellular pathogens that are hidden from antibody and complement. NK cells are large granular lymphocytes that perform innate cytotoxicity. They can begin killing virus-infected cells immediately upon contact, and their killing is enhanced after NK activation by cytokines produced by activated T_H and macrophages. NK cells are not antigen specific. They kill targets when the MHC inhibitory signals are blocked by the infecting virus, either because the virus interferes with MHC expression or the virus peptide on MHC makes it unrecognizable as self. NK cells secrete perforin and granzymes to induce apoptosis in their targets.

    CTL perform adaptive cytotoxicity. They are $CD8^+$ T cells that recognize specific virus peptides synthesized in host cell cytoplasm and presented on MHC class I. CTL must be activated by antigen and co-stimulatory signals from either the infected cell (often dendritic cells which constitutively express B7) or IL-2 from T_H1 cells activated by the same virus. Activated CTL circulate through the body and kill virus-infected cells presenting their specific virus peptide. They also use perforin and granzymes to induce apoptosis in their targets. In addition, CTL can bind targets expressing Fas using FasL and induce apoptosis; this may provide an important immune regulatory mechanism.

    Macrophages eliminate pathogens by engulfing them and killing them with proteolytic enzymes, nitric oxide, and oxygen radicals in the phagolysosome. Macrophages can be activated by binding of T_H1 cell CD40L to their membrane CD40 and by cytokines, especially IFN gamma. Activated macrophages are more effective at killing pathogens such as mycobacteria and leishmania, which inhibit fusion of the lysosome with the phagocytic vesicles or which resist the actions of the killing molecules and continue to live in the macrophage vesicles.

Check Your Performance:

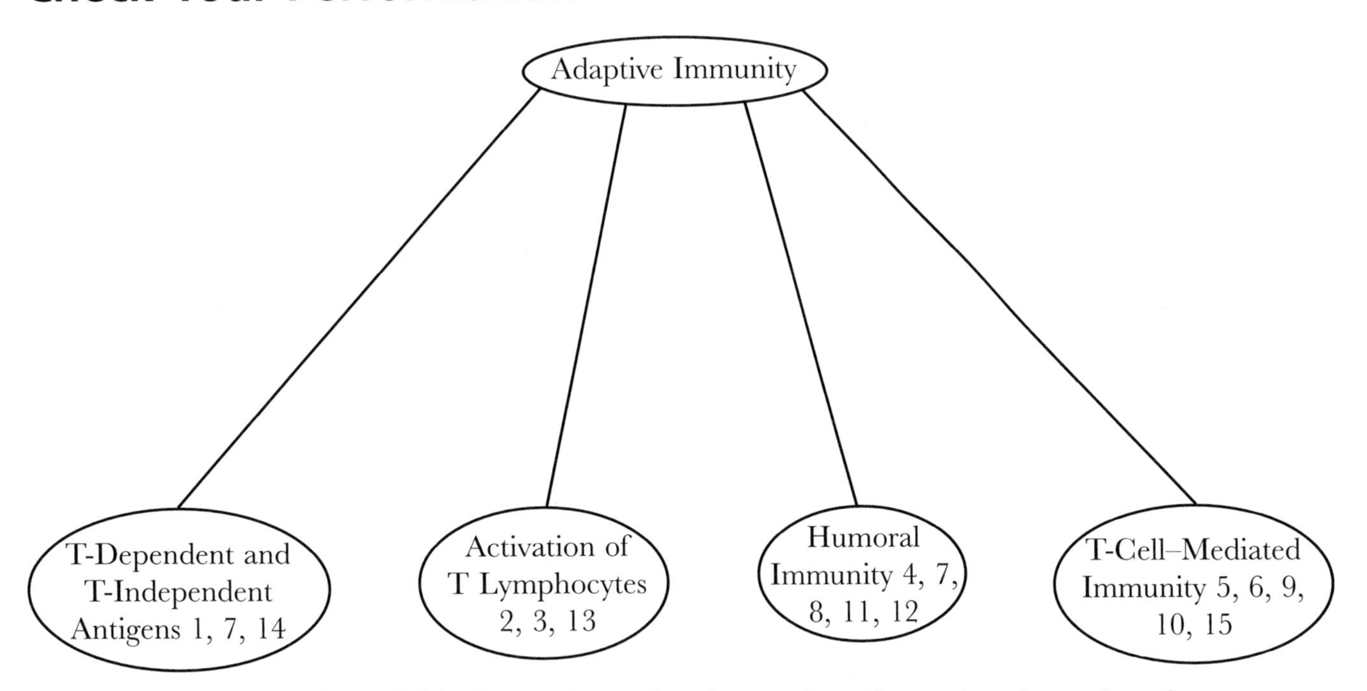

Check your understanding of this chapter by noting the number of questions for each topic you missed on the chapter test.

Unit IV: *Clinical Immunology*

Hypersensitivity and Allergies

Hypersensitivity results from damage done to the body by an immune response. Many immune responses damage the body during antigen removal, causing swelling and pain from inflammation or lysis of virus-infected cells by cytotoxic T lymphocytes (T_C). When the damage is too great, hypersensitivity becomes life threatening. Hypersensitivities are classified into four types based on the mechanism of tissue damage.

ESSENTIAL BACKGROUND

- **Functional classes of antibody (Chapter 3)**
- **Complement activation (Chapter 8)**
- **Inflammation (Chapter 9)**
- **T-cell–mediated immune mechanisms (Chapter 10)**

TOPIC 1: IgE-DEPENDENT HYPERSENSITIVITY (TYPE I)

KEY POINTS

✓ *What is the immune mechanism of type I hypersensitivity?*

✓ *What are some clinical examples of type I hypersensitivity?*

✓ *How is type I hypersensitivity treated?*

Most of what we call "allergy" is **type I** or **immediate hypersensitivity**, mediated by IgE and mast cells. The risk of developing a type I hypersensitivity (**atopy**) is greater if one has atopic parents or higher IgE levels. Serum IgE has C_H2 domains that bind FcεRI on mast cells and basophils. IgE production requires interleukin (IL)-4 production by T_H2 cells. Eosinophils stimulated by antigen–IgE binding to FcεRI provide the most effective immune response to worm parasites, so in locales with high helminth burdens, IgE production is advantageous.

What makes certain molecules **allergens** is not known; grasses, pollens, animal dander, and foods usually reach mucosal surfaces at very low doses. Initial exposure to allergen results in **sensitization**: IgE synthesis and binding to mast cells in smooth muscle, blood vessels, and mucosal linings that is often asymptomatic. Subsequent exposure to antigen cross-links IgE on mast cell FcεRI and immediately triggers allergic symptoms by releasing preformed mast cell molecules stored in cytoplasmic granules. Among these compounds are **histamine**, which causes smooth muscle contraction, mucus release, vasodilatation, and increased capillary permeability; proteolytic enzymes, which break down tissue matrix proteins; and tumor necrosis factor (TNF) alpha, which increases adhesion molecule expression on endothelial cells to promote leukocyte extravasation. Depending on the site of antigen contact, mediator release stimulates mucus production and airway constriction, increased blood flow and movement of fluid into tissues, or gastrointestinal cramping and vomiting. Severe allergic reactions also result in life-threatening systemic anaphylaxis, leading to vascular shock, cardiac arrhythmias, blocked airways, and fluid accumulation in the respiratory system. People whose airways are particularly sensitive to mast cell products develop asthma, where airway constriction blocks expulsion of air and suffocation may occur.

Mast cells stimulated by antigen binding to IgE–FcεRI complexes synthesize another group of mediators that cause symptoms (**late-phase response**) several hours later. These mediators include chemokines and platelet-activating factors to attract leukocytes, cytokines (including IL-4) to activate eosinophils and stimulate their synthesis in the marrow, and leukotrienes (SRS-A, slow reacting substance of anaphylaxis) to promote blood flow, smooth muscle constriction, and mucus secretion.

Type I hypersensitivity is measured by skin testing or by radioallergosorbent test (RAST), a radioimmunoassay using anti-IgE. Common type I hypersensitivities include hay fever, allergic asthma, urticaria (hives), eczema (itchy rash), food allergies, and systemic anaphylaxis to insect stings. The best treatment for allergy is complete allergen avoidance. Medications to prevent or lessen allergic symptoms include sodium cromoglycate to block mast cell degranulation, antihistamines and epinephrine to block mediator actions on the tissues, and immunosuppressive drugs like steroids to block helper T-cell (T_H) activity. Chronic desensitization, administration of allergen beginning with very low doses that are increased over many months, is not effective for every allergen or for every individual, and its mechanism of action is unresolved. Current hypotheses are that the immune response is switched from IgE to IgG, which binds allergen before it can trigger mast cells, or that the treatment induces suppressor T cells that block B-cell activation and IgE synthesis in response to allergen.

Topic Test 1: IgE-Dependent Hypersensitivity (Type I)

True/False

1. A skin test for type I hypersensitivity will be positive only if the person has already been exposed to the allergen.
2. Active allergy symptoms can be treated by desensitizing the sufferer with IgG antibody to the allergen.

Multiple Choice

3. Humans probably make IgE responses because
   a. IgE binds more efficiently to low doses of antigen than IgG.
   b. IgE is protective against dangerous pollens.
   c. IgE triggers eosinophils to release products toxic to helminth parasites.
   d. Their T cells were not properly negatively selected to self-IgE in the thymus.
   e. They cannot produce enough IgG to protect themselves against allergens.

4. Fran walks outside on a beautiful day and takes a deep breath of ragweed pollen, to which she has a strong type I hypersensitivity. Which event below will *not* occur within 30 minutes due to this hypersensitivity?
   a. A local inflammatory response is induced, resulting in a runny or stuffy nose.
   b. IgE specific for ragweed pollen is synthesized by B cells in the local lymph nodes.
   c. Mast cells respond to the antigen-IgE signal by releasing preformed histamine.
   d. Ragweed pollen binds IgE present on mast cell FcεRI in the respiratory tract.
   e. Systemic effects of hypersensitivity such as anaphylactic shock may occur.

Short Answer

5. How can allergens cause such widely varied symptoms as asthma, hives, and nausea?
6. Some people are "allergic" to hot or cold temperatures; they experience swelling in their extremities or difficulty breathing. What is the allergen? Can you propose a mechanism for these allergic attacks?

Topic Test 1: Answers

1. **True.** The test detects sensitized mast cells bearing IgE for the allergen.
2. **False.** Desensitization over many months may induce IgG that will prevent allergic symptoms, but IgG is not used as a therapy for acute allergic symptoms.
3. **c.** IgE's effectiveness in triggering eosinophil and mast cell defenses against helminth infections is probably the selective pressure to maintain our ability to make it.
4. **b.** IgE is synthesized, but even in a memory response it takes a few hours for memory B cells to begin dividing and differentiating into IgE-secreting plasma cells.
5. Symptoms depend on where the mast cells release their mediators. Inflammatory mediators released in the respiratory tract cause constriction of the airways and increased mucus secretion, leading to sneezing and wheezing. Mediators released in the skin cause swelling, reddening, and itching. Mediators released in the gastrointestinal tract cause increased fluid secretion and peristalsis, leading to vomiting and diarrhea.
6. Temperature extremes may induce mast cell degranulation in the absence of allergen or IgE production. The mediators would cause "allergic" symptoms.

TOPIC 2: ANTIBODY-DEPENDENT CYTOTOXICITY (TYPE II)

KEY POINTS

✓ *What is the immune mechanism of type II hypersensitivity?*

✓ *What are some clinical examples of type II hypersensitivity?*

✓ *How is type II hypersensitivity treated?*

Type II hypersensitivity results when IgG or IgM binds cell surface antigen. Membrane antigen–antibody complexes activate complement or bind to FcγR on killer cells that perform antibody-dependent cell-mediated cytotoxicity (ADCC). Both processes result in lysis of the antibody-coated cell. Complement- and ADCC-mediated lysis are normally directed against pathogens and are important protective mechanisms of the immune system.

Clinical examples of type II hypersensitivities include autoimmune diseases in which antibodies are produced to membrane proteins, such as acetylcholine receptor in myasthenia gravis, thyroid hormone receptor in Graves' disease, and erythrocyte membrane proteins in autoimmune hemolytic anemia. In **hyperacute graft rejection**, preformed antibodies to blood group or transplantation antigens cause immediate, severe, and nonreversible damage to the graft. Aspirin, penicillin, and other drugs bind erythrocyte membrane proteins; antibodies to those drugs also bind erythrocytes and damage them. Type II hypersensitivities are detected by agglutination (Chapter 2). Clinical intervention for type II hypersensitivities involves prevention (blood and tissue cross-matching), discontinuing the offending drugs, supportive therapy, and immune suppressants.

Topic Test 2: Antibody-Dependent Cytotoxicity (Type II)

True/False

1. Type II hypersensitivity occurs when antibody binds soluble antigen and causes its deposition on blood cells.
2. Treatment for type II hypersensitivity includes giving anti-inflammatory medications.

Multiple Choice

3. All of the following are type II hypersensitivities *except*
   a. A blood transfusion reaction to AB antigens on erythrocytes.
   b. Autoimmune hemolytic anemia, production of autoantibodies to erythrocytes.
   c. Drug-induced hemolytic anemia, production of antibodies to erythrocyte-binding medications.
   d. Grave's disease, production of autoantibodies to thyroid-stimulating hormone receptor on thyroid cells.
   e. Serum sickness, production of antibodies to foreign antibodies.
4. Type II hypersensitivity results in all of the following *except*
   a. Attraction and activation of inflammatory cells.
   b. Increased vascular permeability.
   c. Lysis of antibody-coated cells by natural killer cells.
   d. Opsonization of microbes for recognition by phagocytic cells.
   e. Release of proteases by macrophages.

Short Answer

5. How are type II hypersensitivities diagnosed?
6. Could one develop both type I and type II hypersensitivity to aspirin?

Topic Test 2: Answers

1. **False.** Type II hypersensitivity occurs when antibody binds to cell surface antigens and results in lysis of the cell.
2. **True.** Treatment for type II hypersensitivity includes giving anti-inflammatory medications to counteract inflammatory complement mediators.
3. **e.** Serum sickness is a type III hypersensitivity because the antigen is soluble, not membrane bound.
4. **d.** Type II hypersensitivity involves cell surface or soluble antigens and usually occurs in the absence of infectious microbes.
5. Type II hypersensitivities are diagnosed by agglutination of cells from the patient with anti-Ig antibodies, indicating that the patient's cells are coated with antibodies.
6. **Yes.** Production of IgE would result in immediate hypersensitivity. IgG binding to aspirin adsorbed onto erythrocyte membranes could trigger complement lysis and type II hypersensitivity.

TOPIC 3: IMMUNE COMPLEX-DEPENDENT HYPERSENSITIVITY (TYPE III)

KEY POINTS

✓ *What is the immune mechanism of type III hypersensitivity?*

✓ *What are some clinical examples of type III hypersensitivity?*

✓ *How is type III hypersensitivity treated?*

Type III hypersensitivity is caused by immune complex deposition in the tissues, where complexes activate the classical complement cascade and cause tissue damage. Immune complexes are normally removed from the circulation by cells bearing complement receptors and FcγR. When antigen persists or high levels of antigen are encountered, immune complexes reach such high levels that they are no longer soluble. Common sites of deposition and tissue damage are the kidneys and joints. Once cells are damaged and an inflammatory response is initiated, release of cytoplasmic enzymes and influx of inflammatory cells prolong the hypersensitivity.

Clinical examples of type III hypersensitivity include skin reddening and necrosis after repeated intradermal antigen injections (**Arthus reaction**), serum sickness in response to passive immunization with foreign antiserum, and occupational diseases in which antibody is produced to soluble environmental antigens that are encountered repeatedly. **Rheumatoid factor** detectable in the serum of patients with rheumatoid arthritis is an IgM anti-IgG antibody thought to contribute to arthritic joint inflammation. Persistent malaria, some virus infections, and leprosy can lead to type III hypersensitivity. Type III hypersensitivities are detected by immunofluorescence of tissue biopsies and precipitation of serum immune complexes. Patients receive supportive therapy until the antigen has been cleared by the antibody, plasmapheresis to reduce immune complex levels, and immunosuppressive therapy.

Topic Test 3: Immune Complex-Dependent Hypersensitivity (Type III)

True/False

1. Type III hypersensitivities are usually caused by transient high levels of antigen–antibody complexes.
2. Type III hypersensitivities may be autoimmune in origin.

Multiple Choice

3. As he cleared brush near his home, Frank was bitten by a rattlesnake. He went to the emergency room for treatment with horse IgG anti-rattlesnake venom. About a week after the treatment, Frank experienced a rash, fever, swollen lymph nodes, and pains in his joints. These symptoms are probably due to
   a. A T-cell memory response to horse IgG.
   b. Cross-reactivity between horse IgG and human IgG.
   c. Late-phase damage caused by the rattlesnake venom.
   d. Production of anti-horse Ig, which triggered a type III hypersensitivity.
   e. Production of anti-rattlesnake venom, which triggered a Type III hypersensitivity.
4. In the situation described in Question 3 above, Frank can be treated with
   a. Antiserum to complement and to block its activation.
   b. Human anti-horse IgG to more quickly clear the horse antibody.
   c. Immunosuppressive drugs to block B-cell production of antibody.
   d. Plasmapheresis to remove antigen–antibody complexes from the blood.
   e. Rattlesnake venom to absorb the horse anti-venom antibody.

Short Answer

5. What organs are most often damaged by type III hypersensitivity?
6. How could you modify horse anti-rattlesnake venom to prevent serum sickness if Frank (Question 3 above) is bitten again by a rattlesnake?

Topic Test 3: Answers

1. **False.** Type III hypersensitivities are often caused by chronic high levels of antigen–antibody complexes, where the antigen is a self-peptide or where a foreign protein or pathogen is chronically present.
2. **True.** Antibody-mediated autoimmune diseases are often type III hypersensitivities.
3. **d.** Frank made antibody to the foreign horse IgG molecules. When complexes of his antibody with horse IgG reached high enough levels they were deposited in the tissues, where they activated complement and caused serum sickness.
4. **d.** Removing the antigen–antibody complexes more quickly than they are being cleared by $Fc\gamma R^+$ cells would alleviate the type III hypersensitivity. The other treatments would likely worsen Frank's condition.

5. The kidneys and joints trap immune complexes and are often damaged.
6. Immunodominant antigens on horse IgG are in C_L and C_H. Fab, humanized, or Fv antivenom would be less immunogenic.

TOPIC 4: DELAYED-TYPE HYPERSENSITIVITY (TYPE IV)

KEY POINTS

✓ *What is the immune mechanism of type IV hypersensitivity?*

✓ *What are some clinical examples of type IV hypersensitivity?*

✓ *How is type IV hypersensitivity treated?*

Delayed-type hypersensitivity (**DTH**) occurs 48 to 72 hours after antigen contact and is mediated by antigen-specific T_H1 cells and activated macrophages. T_H1 cells secrete chemokines to attract macrophages, interferon (IFN) gamma to activate macrophages, TNF alpha and TNF beta to upregulate adhesion molecules on local blood vessels, and IL-3 and granulocyte-macrophage colony-stimulating factor to increase bone marrow output of monocytes. Although macrophages are not antigen specific, they are activated only in the vicinity of antigen-activated T cells. Initial contact with antigen (sensitization) may induce memory T_H1 cell production without symptoms.

Clinical examples of DTH include contact sensitivity to poison ivy, latex, nickel in coins and jewelry, and cleaning products. Nonprotein antigens stimulate T cells by forming complexes with skin proteins; peptides of the altered self-proteins are immunogenic. Skin rashes (exanthems) occur by a similar mechanism to those seen during smallpox, measles, chickenpox, and herpes infections. Tubercles formed in the tissues of patients with tuberculosis are granulomas formed around slow-growing mycobacteria through the influence of T_H1 cells. Allograft rejection and graft-versus-host reactions also use T_H1-activated macrophages.

DTH is measured by skin testing; an example is the tuberculosis skin test. Redness and swelling 2 to 3 days after tuberculin injection indicates prior exposure to *M. tuberculosis*, not necessarily active infection. DTH is treated by antigen avoidance and by immune suppressants.

Topic Test 4: Delayed-Type Hypersensitivity (Type IV)

True/False

1. DTH symptoms are maximal at 48 to 72 hours after contact with antigen.
2. DTH T cells are not normally involved in cellular immunity to pathogens.

Multiple Choice

3. DTH
   a. Can be passively transferred with $CD4^+$ T cells.
   b. Causes chickenpox.
   c. Involves cell damage by cytotoxic T cells.
   d. Is mediated by memory macrophages.
   e. Occurs 1 to 2 weeks after antigen exposure.

4. During delayed-type hypersensitivity macrophages
   a. Are not antigen specific.
   b. Are stimulated by IFN gamma.
   c. Do not depend on antibody for antigen recognition.
   d. Kill neighboring cells, whether infected or not.
   e. All of the above

Short Answer

5. If memory T_H1 cells are already present, why does DTH take 2 to 3 days to cause symptoms?
6. How can T cells be activated by nonprotein molecules like nickel?

Topic Test 4: Answers

1. **True.** It takes 48 to 72 hours after contact with antigen for T_H1 cells and effector macrophages to reach the antigen and become fully activated.
2. **False.** DTH T cells (T_H1) are normally involved in cellular immunity to intravesicular bacteria and parasites.
3. **a.** DTH is an example of cellular immunity, which can be transferred with specific T cells.
4. **e.** The antigen specificity in DTH comes from the T cells, not the macrophages. Antibodies are not involved in DTH.
5. Memory T_H1 cells are present in the body but must reach the site of antigen in significant numbers to be activated by antigen to produce inflammatory cytokines, recruit macrophages, and cause fluid accumulation, which result in a visible skin reaction.
6. Nonprotein molecules like nickel bind tissue proteins and act as haptens. T_H1 cells respond to altered self peptides.

CLINICAL CORRELATION: BAD BLOOD

Susan was given a transfusion of three units of whole blood after a difficult delivery. Shortly after the third unit was transfused, Susan experienced fever, chills, nausea, and hemoglobin in her urine. A quick check of the blood cross-matching showed that one unit of type AB negative blood had been given to Susan, who is type A negative. Susan was given supportive therapy and recovered without further incident. A and B erythrocyte antigens are complex carbohydrates on membrane glycolipids. Susan, with type A blood, has A antigen. Type B erythrocytes have B antigen, type AB have both A and B antigens, and type O have neither antigen. A and B carbohydrates are also present on intestinal bacteria; humans make IgM antibodies to bacterial antigens not shared with their own erythrocytes. When Susan was given AB blood, her preformed IgM anti-B bound the donated red blood cells, initiating a type II hypersensitivity.

DEMONSTRATION PROBLEM

Susan's daughter Sara was pale and her lips and fingernail beds were blue several minutes after birth. Her erythrocyte numbers were only 80% of normal, her blood type was AB positive, and her erythrocytes agglutinated with anti-human IgG. Based on these results, the pediatrician diagnosed **hemolytic disease of the newborn** (**erythroblastosis fetalis**). Sara was given supportive oxygen and an exchange transfusion of AB negative erythrocytes. Erythrocytes have a membrane protein called Rh (rhesus) factor. Hemolytic disease of the newborn occurs when the mother is Rh^- (her cells do not have the protein) and the baby is Rh^+. If some of the baby's blood enters the mother's circulation, the mother makes an IgG anti-Rh that crosses the placenta and activates complement to lyse baby's red blood cells. Hemolytic disease of the newborn does not usually occur with first pregnancies, but each exposure to fetal Rh^+ erythrocytes increases production of IgG anti-Rh. Once Sara tested Rh^+, Susan was given an injection of **Rhogam**, human anti-Rh antibody, to prevent future children from being born anemic.

Why was Sara given Rh negative blood, when her own type was Rh positive? Why didn't Susan's anti-B antibodies harm Sara's AB^+ erythrocytes? How does Rhogam prevent Susan from making more anti-Rh antibodies?

SOLUTION

Because the anemia was caused by maternal anti-Rh antibody, transfused Rh^+ cells would also be lysed; Rh^- cells would not. There are no "natural" antibodies to Rh so Sara will not have a type II hypersensitivity to Rh^- cells. Susan's anti-B antibodies were IgM because A and B antigens are T independent; IgM does not cross the placenta. Rh is a protein antigen and generates an IgG response. Rhogam prevents Susan from making more anti-Rh antibodies by binding Sara's erythrocytes in Susan's circulation and destroying them before they stimulate IgG production.

Chapter Test

True/False

1. Hypersensitivity reactions involve immune responses not seen against common pathogens.
2. Type II and type III hypersensitivity are mediated predominantly by IgG.
3. Only a portion of the population is atopic because not everyone can make IgE.
4. Type III hypersensitivity is diagnosed by the agglutination of blood cells with anti-human IgG.
5. Inflammation is a process that occurs in all four hypersensitivity types.

Multiple Choice

Classify each of the conditions in 6 to 10 below as one of these hypersensitivities:

a. Type I
b. Type II
c. Type III
d. Type IV

6. Allograft rejection: cytotoxic T lymphocytes and macrophages activated by T_H1 cells kill transplanted cells.
7. Drug-induced hemolytic anemia: medications bind to red blood cells, anti-drug antibodies are made that damage cells.
8. Farmer's lung: IgG antibody to soluble environmental antigen forms immune complexes in alveolar tissue.
9. Runny nose every time rye grass blossoms.
10. Vomiting and diarrhea from eating a cookie containing wheat flour.

Short Answer

11. How are skin testing for type I and type IV hypersensitivity different?
12. Both type II and type III hypersensitivities are caused by IgG antibodies; how are they different?
13. Type II and type IV hypersensitivity both involve cytotoxicity by macrophages; how are they different?
14. How is desensitization in the case of type I hypersensitivity different from giving Rhogam to prevent erythroblastosis fetalis?

Essay

15. Why are T_H1 cells important in the body's immune response to *Mycobacterium tuberculosis*, and when and why does this reactivity become a hypersensitivity?

Chapter Test: Answers

1. **F** 2. **T** 3. **F** 4. **F** 5. **T** 6. **d** 7. **b** 8. **c** 9. **a** 10. **a**

11. Intradermally injected antigen induces redness and swelling in 15 to 30 minutes in type I hypersensitivity and in 48 to 72 hours in DTH (type IV).
12. In type II hypersensitivity antigen is on cell membranes and those cells are killed by complement or ADCC. In type III hypersensitivity antigen is soluble; damage occurs when antigen–antibody complexes become insoluble and activate complement where they are deposited in blood vessels, joints, and kidneys.
13. In type II hypersensitivity, macrophages bind and kill antibody-coated cells. In type IV hypersensitivity, T_H1 cells bind antigen and release cytokines to signal nearby macrophages to kill; the target cells may not have foreign antigen on their membranes. Macrophage cytotoxic mechanisms are the same for both hypersensitivities.
14. Desensitization involves giving low doses of antigen to induce suppressor T cells or a non-atopic antibody isotype. Rhogam is passive antibody used to prevent erythroblastosis fetalis by removing antigen.
15. *M. tuberculosis* prevents macrophages that engulf it from fusing their lysosomes with the phagosomes. T_H1 cells activated by *M. tuberculosis* peptides on macrophage class II secrete

IFN gamma to stimulate lysosome–phagosome fusion, NO and O_2 cytotoxicity, and membrane expression of MHC class II. Activated macrophages also secrete IL-12 to activate T_H1 inflammatory cells. If this immune response is successful, the pathogen is eliminated and the response ends. If it is not successful, activated macrophages kill surrounding uninfected tissue in their attempts to remove the pathogen and cause DTH.

Check Your Performance:

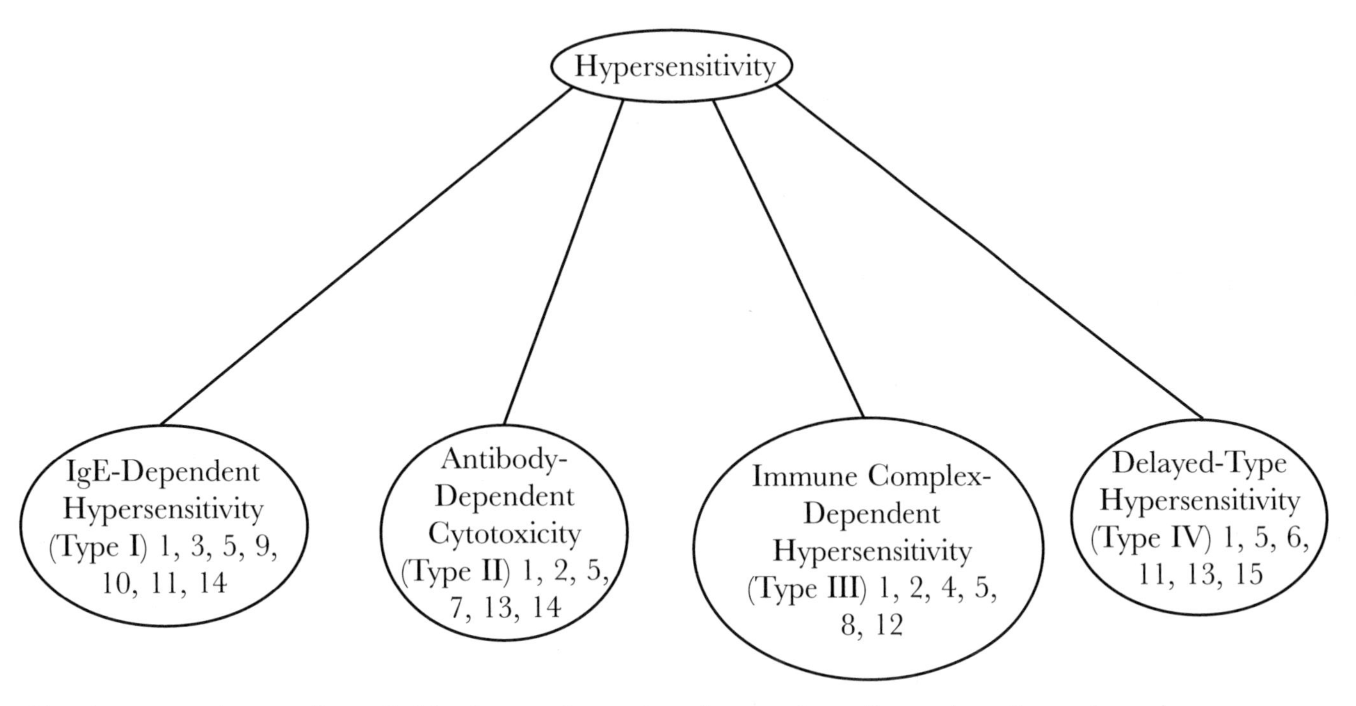

Check your understanding of this chapter by noting the number of questions for each topic you missed on the chapter test.

Autoimmunity and Immune Deficiencies

Immune responses are heavily regulated. Despite this regulation and mechanisms for clonal deletion of self-reactive lymphocytes, the immune system occasionally attacks self. At the other end of the immune spectrum, immune deficiencies result in life-threatening infections.

ESSENTIAL BACKGROUND

- **Measuring immunity (Chapter 2)**
- **T-cell education (Chapter 6)**
- **Complement activation (Chapter 8)**
- **Innate and adaptive immune mechanisms (Chapters 9 and 10)**
- **Hypersensitivities (Chapter 11)**

TOPIC 1: IMMUNE SYSTEM REGULATION

KEY POINTS

✓ *How are immune responses regulated?*

✓ *How is tolerance achieved and maintained?*

✓ *How can tolerance be broken or circumvented?*

The basis of a functional immune system is genetic: the ability to make mature lymphocytes and accessory cells bearing the required antigen-specific receptors, co-receptors, major histocompatibility complex (MHC), and co-stimulatory molecules. Functional genes for complement, inflammatory mediators, and cytokines are also required. Nutrition, the presence of chronic disease, and lifestyle all affect immune responsiveness. As we age, our immune systems work less efficiently; thymus atrophy results in fewer circulating and functional T cells, weaker primary (but not memory) responses, and more infections, cancer, hypersensitivity, and autoimmunity.

Immune responses are regulated by antigen, antibody, cytokines, and hormones. Common bacterial antigens activate complement and stimulate macrophages to express co-stimulatory molecules. Antigen stimulates adaptive immunity by activating lymphocytes, which in turn make antibody to activate complement and cytokines to increase antigen elimination and recruit additional leukocytes. As antigen is eliminated, activation ceases; high affinity B cells are selected by low antigen concentrations. Antigen also negatively regulates immunity; when antigen binds immature cells or binds without co-stimulatory signal, lymphocytes are killed (**clonal deletion**) or inactivated (**clonal anergy**). Secreted antibodies compete with B cells for antigen. Immune complexes bind Fc receptor on B cells; at low levels they increase the sensitivity of B cells to antigen and increase responsiveness, whereas at high levels they give negative signals to B cells to reduce humoral responses. Anti-idiotype antibodies bind secreted and membrane antibodies to inhibit humoral immunity. T_H1 and T_H2 cytokines inhibit one another's production and function: T_H1 cells stimulate cellular immunity and suppress humoral immunity, whereas T_H2 cytokines have the opposite effect. Interleukin (IL)-1 induces sleep and fever and stimulates release of pituitary hormones [corticotropin, thyroid-stimulating hormone (TSH), and growth hormone]. Thymic, steroid, and peptide hormones also influence our ability to make immune responses.

Natural tolerance occurs for a number of reasons. **Central tolerance** results from clonal deletion of self reactive T and B cells during development. The particular human leukocyte antigen (HLA) alleles available to present self antigen to developing T cells influence which peptides are presented well enough to induce clonal deletion. Clonal anergy occurs in the periphery, when immature B cells encounter antigen that extensively cross-links B-bell receptor (BCR) and T cells encounter unprocessed antigen or processed antigen in the absence of co-stimulatory signals. Clonal anergy maintains tolerance to self antigen not seen in the thymus and marrow. **Immunological ignorance** due to antigen sequestration is another reason for lack of immune responses; for example, T and B cells are not exposed to the antigen in the lens of the eye (which has no circulation), sperm cells, and neural cells. **Suppressor CD8$^+$ T cells** reduce both humoral and cellular immune responses. Suppression is transferable with T cells and is antigen specific, but suppressor T cells are less well characterized than other T cells and may actually be cytotoxic T lymphocytes (T_C), T_H1, or T_H2 cells producing suppressive cytokines.

Experimental tolerance is induced by antigen exposure in an immature animal or an immature lymphoid system, exposure to weak immunogens such as soluble nonaggregated proteins and nonmetabolizable substances, very low (T cells) or very high (T and B cells) antigen doses, and persistent antigen stimuli. T cells are easier to make tolerant and remain tolerant longer than B cells; deleted clones are not replaced in adult animals. Tolerance, including self tolerance, can be "broken" or circumvented. If epitopes are presented on new carrier molecules, T_H cells may help B cells (not clonally deleted) respond to antigen. Infection with organisms bearing antigens cross-reactive with self is known to initiate several autoimmune diseases in humans.

Topic Test 1: Immune System Regulation

True/False

1. Depending on their concentration, antigen–antibody complexes can stimulate or inhibit B cell activation.

2. It is easier to make T cells tolerant than to make B cells tolerant.

Multiple Choice

3. Which of the following does *not* play a role in regulation of immune responsiveness?
   a. Antibody binding to antigen blocks antigen binding to B cells.
   b. Declining antigen levels result in selection of high-affinity B cells.
   c. High levels of immune complexes inhibit B-cell activation.

d. Increased antigen levels induce the activation of more T cells in the thymus.
e. T_H1 cytokines tend to inhibit humoral immunity.

4. Tolerance is *not* due to
   a. Administration of antigen with adjuvant.
   b. Clonal anergy of self reactive mature T cells that bind antigen without co-stimulation.
   c. Clonal deletion of self reactive immature B cells whose BCR is extensively cross-linked by antigen.
   d. Failure of particular MHC alleles to present certain peptides.
   e. Lack of access of the immune system to antigens in the eye.

Short Answer

5. How is immunological ignorance related to tolerance?
6. How might an infection "break" tolerance?

Topic Test 1: Answers

1. **True.** At low concentration, antigen–antibody complexes make B cells more sensitive to antigen; at high concentration, they inhibit B-cell activation.
2. **True.** T cells become unresponsive at both very low and very high antigen concentrations, whereas B cells usually undergo only high dose tolerance.
3. **d.** Antigen does not directly influence the amount of T-cell development.
4. **a.** Adjuvants generally stimulate immunity.
5. Immunological ignorance occurs because the immune system has no contact with antigen. If contact is made due to infection or injury, an autoimmune response might occur.
6. An infection may "break" tolerance by stimulating the immune system so strongly with an antigen that resembles self that normally tolerant T cells respond and then go on to damage self tissue.

TOPIC 2: AUTOIMMUNE DISEASES

KEY POINTS

✓ *How might MHC alleles influence the appearance of autoimmune disease?*

✓ *How can anti-receptor antibodies either increase or decrease cell function?*

✓ *What are some antibody-mediated and T-cell–mediated autoimmune diseases?*

✓ *Why are many immunosuppressive therapies aimed at inhibiting T-cell function?*

Autoimmunity is caused by adaptive immune responses against "self" antigen. Because self antigens are continually present in the body, tissue damage can be life threatening. Risk factors for autoimmune disease include the presence of certain HLA alleles, sex hormone levels, and infection. Autoimmune diseases can be caused by antibodies or T cells and may be targeted to specific tissues or be systemic.

Different alleles of HLA bind peptide differently. Two possible models attempt to explain HLA linkage to autoimmunity. One scenario is that certain HLA alleles are better at presenting self peptides that cross-react with pathogen peptides to mature T cells. The other scenario proposes that certain HLA alleles are less efficient at presenting self peptides to developing T cells in the thymus, so that negative selection fails. These scenarios are illustrated by two specific examples of HLA involvement in autoimmune disease: insulin-dependent diabetes mellitus (IDDM) and ankylosing spondylitis. HLA-DQ beta chain normally has an aspartic acid at position 57, which forms a salt bridge with the alpha chain. People with an amino acid that does not form a salt bridge at position 57 are at increased risk for IDDM. Other class II alleles are protective for IDDM, such that people with those alleles have a greatly reduced chance of developing IDDM. HLA-B27 is associated with an 80-fold increased risk of alkylosing spondylitis, characterized by spinal inflammation. Recent research shows that B27 binds peptides in the absence of tapasin, advantageous in virus infections where the virus interferes with tapasin production. Peptides bound in the absence of tapasin are different from those bound in its presence and may include self peptides.

When autoimmune disease is caused by autoantibodies, the antigen can often be identified and the disease mechanism classified as type II or type III hypersensitivity. In autoimmune hemolytic anemia, antibodies to erythrocyte antigens initiate complement lysis and phagocytosis in the spleen reticuloendothelial system. Antibodies to platelets and neutrophils also cause their depletion. Treatment for these autoimmune diseases involves removal of the spleen. Complement activation at levels too low to lyse cells can still cause damaging inflammation. Antibody-dependent cell-mediated cytotoxicity (ADCC) also causes tissue damage and cell death. An example of antibody-mediated autoimmunity is Hashimoto's thyroiditis, where antibodies to thyroid enzymes or hormones damage the thyroid.

Antibodies to receptors may stimulate or inhibit receptor function. Antibodies to TSH receptor in the thyroid stimulate thyroid hormone production and induce hyperthyroidism (overactive thyroid) in Graves' disease. Antibodies to the receptor for acetylcholine, a neurotransmitter, block signals to muscle cells and cause progressive muscle weakness in myasthenia gravis. Antibodies to insulin receptor may mimic insulin function and cause low blood sugar or block insulin function and cause high blood sugar. Antibodies to connective tissue collagen damage the kidney in Goodpasture's syndrome. In systemic lupus erythematosis, IgG production to many self antigens damages tissue throughout the body (systemic disease). Lyme arthritis seen in some people infected with *Borrelia burgdorferi*, the spirochete that causes Lyme disease, results from antibody production to the persisting pathogen.

T-cell–mediated damage results in several autoimmune diseases, including multiple sclerosis, rheumatoid arthritis, and insulin-dependent diabetes. It is more difficult to identify autoimmune T cells and the antigen to which they are responding than it is to identify antibodies and their antigens. The specificity of autoimmune T cells may be identifiable in animal models of disease, such as experimental autoimmune encephalitis, a model for multiple sclerosis. Damage in IDDM results from cytotoxic T lymphocytes (CTL) specific for pancreatic beta cell peptides presented on membrane MHC class I.

Because autoimmune disease is mediated by normal immune mechanisms, controlling it without making the patient susceptible to infection is the greatest challenge. Antibodies to lymphocytes, such as broadly specific horse anti-lymphocyte globulin or T-cell–specific monoclonal anti-CD3, are effective in suppressing autoimmunity and transplant rejection. However, they kill lymphocytes regardless of antigenic specificity or activation status and induce immune responses against their nonhuman isotypes that lessens their effectiveness and causes serum sickness. Prednisone is an analogue of the corticosteroid cortisol; it suppresses expression of genes for many cytokines, nitric oxide synthetase, vascular adhesion molecules, and inflammatory mediators like prostaglandins and leukotrienes to reduce overall inflammation. Azathioprine is a cytotoxic drug that blocks DNA synthesis, targeting its suppression to activated lymphocytes and other rapidly dividing cells. Cyclosporin A and FK506 (tacrolimus) bind T-cell cytoplasmic molecules called **immunophilins**, which in turn bind calcineurin and block activation of NF-AT and synthesis of IL-2 during T-cell activation. Doses of cyclosporin A and FK506 can be adjusted to preserve immune function against pathogens while reducing autoimmune damage. However, they are both toxic to the kidney and expensive to administer.

Topic Test 2: Autoimmune Diseases

True/False

1. HLA-B27, an allele associated with greatly increased risk of alkylosing spondylitis, presents unprocessed bacterial proteins to T cells.
2. Autoantibodies are the primary cause of autoimmune disease.

Multiple Choice

3. Autoimmunity probably does *not* result from
   a. Cross-reactivity of pathogen and self antigens.
   b. Expression of self antigen in the thymus or bone marrow.
   c. Failure of MHC to present self peptides to self-specific T cells in the thymus.
   d. Somatic mutations in the BCR antigen-recognition site.
   e. Tissue injury that releases normally hidden self antigens.
4. Possible immunotherapeutic strategy for treating autoimmunity would *not* include administration of
   a. Adjuvants to induce tolerance in antigen-presenting cells (APC).
   b. Antibodies to CD3.
   c. Azathioprine to block T-cell division.
   d. Cyclosporin A to block IL-2 production during T-cell activation.
   e. Prednisone to reduce inflammation.

Short Answer

5. What are the possible reasons that the incidence of autoimmune disease is genetically linked to MHC alleles?
6. How can infections trigger autoimmunity?

Topic Test 2: Answers

1. **False.** HLA-B27 can be loaded with antigen peptide in the absence of tapasin, and some of the peptides may be different from those which usually bind HLA B27. Unprocessed bacterial proteins would not be presented by B27.

2. **False.** Many autoimmune diseases are caused by self-specific T cells, but their antigen specificity is more difficult to determine.

3. **b.** Autoimmunity probably does *not* result from expression of self antigen in the thymus or bone marrow, which should lead to effective clonal deletion of T and B cells specific for that antigen.

4. **a.** Adjuvants activate APC to express more co-stimulatory molecules; this is what might have initiated autoimmunity.

5. Autoimmunity may be genetically linked to MHC because not all alleles present the antigen equally well to T cells. Overpresentation of cross-reactive pathogen alleles or underpresentation of self alleles during lymphocyte development could both result in increased likelihood of autoimmune disease. Activation of T cells is crucial for both cell-mediated and antibody-mediated (IgG) autoimmunity.

6. Infection might trigger autoimmunity by antigenic mimicry (inducing immunity to a cross-reactive epitope), by inducing co-stimulatory molecule expression on APC presenting self peptides to self-specific T cells not deleted in the thymus, or by inducing persistent inflammation.

TOPIC 3: IMMUNE DEFICIENCIES

KEY POINTS

✓ *What are the consequences of innate immune system failure?*

✓ *What are the consequences of adaptive immune system failure?*

✓ *How do pathogens evade the immune system?*

Congenital deficiencies in innate and adaptive immunity have contributed greatly to our understanding of the immune system. Complement deficiencies result in increased bacterial infections, especially with *Neisseria*. Neutropenia is characterized by recurrent and prolonged skin and mucous membrane infections (often staphylococcal) that have minimal clinical signs and respond poorly to antibiotics. If respiratory burst enzymes are defective, microbes grow in phagocytes, causing multiple abscesses and giant-cell granulomas. Adhesion-molecule deficiencies lead to skin infections, gingivitis, and deep tissue abscesses. Natural killer (NK) cell deficiency, seen in Chediak-Higashi syndrome, is associated with increases in frequency of tumors and viral infections (especially herpes viruses).

Humoral, cellular, and combined immunodeficiencies have been characterized; several have been described in clinical correlations in this text. Failure to express transporter associated with antigen processing (TAP) proteins or MHC class I alleles inhibits development of T_C cells and increases susceptibility to virus infections. Failure to express class II alleles inhibits development of T_H cells and increases susceptibility to all pathogens. Bruton's X-linked agammaglobulinemia results from a failure of Btk tyrosine kinase function and blocks B-cell development and antibody production to increase susceptibility to extracellular bacteria. Deficient IgA production increases the incidence and severity of respiratory infections. Antibody deficiencies are the easiest to treat, because infusion of human gamma globulin effectively reduces incidence of infection. DiGeorge syndrome is a failure of the thymus to develop and produce mature T cells; general immunosuppression is the result. Severe combined immune deficiency (SCID) results from lack of a number of different enzymes required for lymphocyte development; infants born with SCID die from infection in the first year of life unless they receive a successful bone marrow transplant. Mutations in Fas generally result in increased incidence of lymphoid cancers because T and B cells do not undergo Fas-mediated apoptosis.

Infectious organisms have evolved many ways of evading their host's immune system. One of the most common is **antigenic variation**, where immunity to one variant is not protective against another. Two very successful examples of antigenic variation are influenza virus and human immunodeficiency virus (HIV). Both are RNA viruses whose error-prone RNA-copying enzymes result in antigenically distinct mutants. Influenza undergoes small changes each year (**antigenic drift**), and the changing serotypes require new vaccines. Occasionally, two influenza viruses from different species co-infect an animal, usually a pig or a fowl, and exchange RNA; when this happens, large **antigenic shifts** occur that can result in pandemics of influenza like the one in 1918. The parasite *Trypanosoma cruzi* undergoes programmed antigen variation; as the host makes antibodies to one surface antigen, that antigen is quickly replaced by expression of a different antigen. Viruses remain invisible to the T cell by becoming latent in the host cell, not replicating and therefore not synthesizing viral proteins that can be processed and presented on MHC class I. Other pathogens avoid macrophage elimination by covering themselves with a phagocytosis-resistant capsule, blocking lysosome fusion with the phagocytic vesicle, or escaping the phagolysosome to live in their own vesicles. Endogenous pathogens avoid antibody and complement by remaining inside host cells. Exogenous pathogens resist B-cell recognition by coating their surface with host proteins and block antibody function by making surface molecules that bind IgG Fc regions or enzymatically cleave IgA. Pathogens also secrete immunosuppressive cytokines.

Topic Test 3: Immune Deficiencies

True/False

1. Humoral immune deficiencies usually result in increased susceptibility to virus infections.
2. Deficiencies in innate immune functions do not have serious consequences because adaptive immunity can take over.

Multiple Choice

3. If MHC class II is not expressed in the thymus, the resulting immune deficiencies would *not* include reduced
   a. Alternative complement activation.
   b. $CD8^+$ T-cell–mediated cytotoxicity.
   c. Macrophage activation to vesicular pathogens.
   d. IgG synthesis.
   e. Recruitment of leukocytes from the marrow.
4. Pathogens evade immune detection by
   a. Altering their antigens periodically.
   b. Binding the Fc region of IgG to block complement activation and opsonization.
   c. Degrading IgA directed against them.
   d. Living in the cytoplasm of host cells.
   e. All of the above

Short Answer

5. What causes influenza pandemics?
6. What animal model is similar to DiGeorge syndrome?

Topic Test 3: Answers

1. **False.** Humoral immune deficiencies usually result in increased susceptibility to bacterial infections.
2. **False.** Deficiencies in innate immune functions can impair both cellular and humoral immunity because of the importance of accessory cells and molecules in adaptive immunity.
3. **a.** If class II is not expressed in the thymus, the resulting immune deficiencies would *not* include reduced alternative complement activation. CTL activity would be reduced due to lack of T_H1 cells.
4. **e.** Pathogens evade immune detection by all of the above mechanisms.
5. Two influenza viruses from different species can exchange genomes and undergo major changes in antigenic specificity (antigenic shift), producing surface antigens to which most members of the human population have no immunity.
6. Nude mice, like children with DiGeorge syndrome, lack a thymus and have impaired humoral and cellular immunity.

CLINICAL CORRELATION: HIV AND AIDS

Acquired immune deficiency syndrome (AIDS) is caused by the human immunodeficiency virus (HIV-1), a retrovirus. HIV is transmitted by body fluid or infected cells. HIV gp120 attaches to CD4 on T_H cells or macrophages, and gp41 promotes fusion of HIV and host cell membranes in the presence of host cell chemokine receptor. HIV is macrophage tropic, preferentially infecting macrophages with CCR5 chemokine receptor, or T-cell tropic, infecting T_H cells with CXCR4. Once in the host cell cytoplasm, HIV **reverse transcriptase** makes a double-stranded DNA copy of its genome, which integrates into host cell DNA. Activation of T_H cells to make an immune response results in production of transcription factor NFκB, which transcribes HIV DNA and initiates virus replication. Anti-HIV antibodies are made that do not block infection but can be used for diagnosis. The initial screening test for HIV is an ELISA followed by a confirmatory Western blot.

HIV kills T cells directly by infection and cytopathic effects and by blocking T-cell activation via gp120 blocking surface CD4 and reduction of IL-2, IL-2R, and T-cell receptor (TCR) expression. HIV binding to CD4 in the absence of other signals induces apoptosis, and membrane fusion of T cells expressing viral gp120 and CD4 results in formation of **syncytia** and cell death. NK cells and HIV-specific CTL lyse infected $CD4^+$ cells. There is some evidence that HIV acts as a superantigen to deplete T cells bearing certain Vβ TCR subsets. Although the immune system responds to HIV with humoral and cellular immunity, both lead to increased T-cell infection and death. Helper T cells are vital for the functioning of many other cells in the immune system; loss of immune function makes the body susceptible to opportunistic infection by normally nonpathogenic organisms.

Those infected with HIV may be asymptomatic for years before developing AIDS. Lymph node follicular dendritic cells trap HIV and nearby T cells become infected, so numbers of infected circulating $CD4^+$ cells and free HIV virions are low. The hematopoietic system replaces destroyed T cells, keeping $CD4^+$ T cell counts in the normal range. Later in infection, replicating virus disrupts the lymph node architecture; more infected T cells and free virions appear in the circulation and the marrow's ability to replace damaged cells declines. $CD4^+$ T-cell numbers decline and opportunistic infections increase.

DEMONSTRATION PROBLEM

Name two ways in which each of the immune deficiencies listed below affects immunity. For example, "No thymus" results in lack of mature T cells in the circulation and secondary lymphoid organs. Increased infections with viruses and vesicular pathogens occur, and B cells cannot produce IgG.

a. No functional C_H alpha immunoglobulin gene.
b. No functional complement C3.
c. No functional MHC invariant chain (Ii).
d. No functional CD8.
e. No RAG-1 or RAG-2.

SOLUTION

a. B cells cannot make IgA; mucosal infections (especially respiratory) increase.
b. No complement activation occurs because both classical and alternative complement pathways use C3. Inflammation is reduced; antigen presentation to T_H cells suffers and both cellular and humoral immunity are impaired. Complement-mediated virus neutralization and pathogen lysis does not occur. In general, infections with *Neisseria* species are increased most in complement deficiencies.
c. Invariant chain blocks endogenous antigen presentation on MHC class II. Class II might present and generate antibodies to more endogenous antigens and fewer exogenous antigens, or class II might be degraded before it could present antigen.
d. CD8 is required for T cells to bind MHC class I–peptide complexes; positive selection of $CD8^+$ T cells cannot occur. Cellular immunity to viruses is severely impaired; cellular immunity to vesicular pathogens and humoral immunity is not affected.
e. Both T- and B-cell development depends on these enzymes for DNA recombination and maturation of lymphocytes. RAG deficiencies result in SCID.

Chapter Test

True/False

1. Microbes that evade cellular immunity by decreasing MHC class I expression on the host cells they infect risk increased detection by NK cells.

2. Autoimmunity is usually initiated by a mutated self antigen.

3. One way we know that steroid hormone levels influence the immune system is that women generally become autoimmune more often than men.

4. Autoimmune hemolytic anemia, caused by antibodies produced against erythrocyte antigens, can be treated by removing the spleen.

5. Prednisone, a cortisol analogue, is used to treat autoimmune reactions because it blocks T-cell activation to self antigens.

Multiple Choice

6. An inability to produce Fas would result in impairment of
   a. Antibody synthesis.
   b. CTL-mediated lysis.
   c. Inflammation.
   d. Macrophage activation by $CD4^+$ T cells.
   e. Natural killing.

7. The process of inducing T-cell tolerance by antigen binding in the absence of co-stimulation is called
   a. Autoimmunity.
   b. Clonal anergy.
   c. Clonal selection.
   d. Negative selection.
   e. Type IV hypersensitivity.

8. An immunologically privileged site would be the
   a. Cornea.
   b. Heart.
   c. Kidneys.
   d. Spleen.
   e. Thymus.

9. HIV can kill $CD4^+$ T cells by
   a. Diverting all their protein synthesis into virus production.
   b. Expressing gp120 on T-cell membranes to make them targets for ADCC.
   c. Inducing apoptosis by binding to CD4 without being presented on MHC II.
   d. Inducing their lysis by HIV-specific CTL.
   e. All of the above probably occur.

10. An autoimmune disease mediated by type III hypersensitivity is
    a. Autoimmune hemolytic anemia caused by anti-A blood group antigen.
    b. Grave's disease caused by anti-TSH receptor antibodies.
    c. IDDM caused by antibodies to blood glucose.
    d. Lyme arthritis caused by antibodies to a persistent infection with *B. burgdorferi*.
    e. Myasthenia gravis, caused by antibodies to the acetylcholine receptor.

Short Answer

11. How does one's MHC genotype influence one's ability to become immune or tolerant to a particular antigen?
12. How can an anti-receptor antibody either promote or inhibit receptor function?
13. What conditions favor induction of tolerance instead of immunity?
14. How can microbes evade humoral immunity?

Essay

15. How can HIV infection cause such profound deficiencies in the immune system?

Chapter Test: Answers

1. **T** 2. **F** 3. **T** 4. **T** 5. **F** 6. **b** 7. **b** 8. **a** 9. **e** 10. **d**

11. Each MHC protein presents some peptides very well, some less well, and others not at all. A self antigen that cannot be presented with high enough affinity to induce clonal deletion will be recognized by mature T cells but will probably not activate them efficiently enough to cause disease. If a pathogen stimulates overpresentation of that self antigen, alters the self so it is presented more strongly, or activates those self-specific T cells because they bind tightly to a cross-reactive epitope on the pathogen, autoimmunity can result. If the self protein were presented well in the thymus, clonal deletion would occur and no self-specific T cells would be present to react to the pathogen antigen.
12. Some antibodies to receptors bind in a way that mimics binding of the normal ligand, resulting in activation. Other antibodies bind in a way that blocks ligand binding without activating the cell, so inhibition is the result.
13. Tolerance induction is favored by exposure of immature cells to antigen, exposure to weak immunogens such as soluble nonaggregated proteins and nonmetabolizable substances, very low (for T cells) or very high (for both T and B cells) antigen doses, and persistent antigen stimuli.
14. Microbes evade humoral immunity with capsules that resist phagocytosis or by blocking lysosome–phagosome fusion, living inside cells without expressing their proteins on the plasma membrane, antigenic variation so that antibodies lose their effectiveness, coating themselves in host proteins or proteins that bind antibody Fc regions, and secreting enzymes that degrade antibodies or cytokines that inhibit humoral immune effector cells.
15. HIV infects and directly kills $CD4^+$ cells by substituting virus production for synthesis of proteins needed for cellular metabolism. HIV also kills by binding CD4 and, in the absence of antigen and co-stimulation, inducing apoptosis or by expressing gp120 on T-cell membranes to bind CD4 on other T cells and form syncytia. Antibodies made against gp120 and gp41 bind infected T cells expressing those antigens and initiate complement-mediated or ADCC-mediated killing. CTL specific for HIV peptides and NK cells detect virus-infected cells and kill them. T-cell cytokines are required for activation of B cells to perform isotype switching and produce memory cells and for macrophage and CTL activation against endogenous pathogens. Helper cell cytokines also regulate immune responses and induce production of additional leukocytes in the marrow.

Check Your Performance:

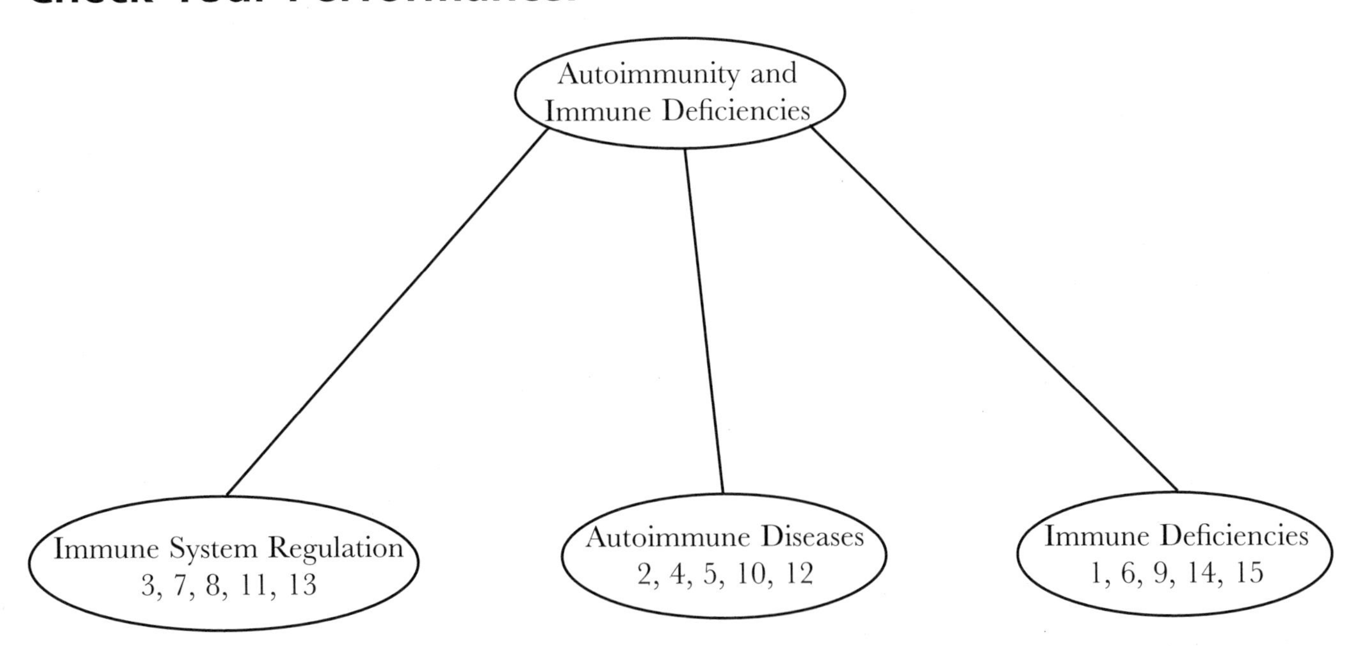

Check your understanding of this chapter by noting the number of questions for each topic you missed on the chapter test.

Manipulating the Immune System

Humans first manipulated the immune system by deliberately exposing themselves to infectious organisms under conditions where they hoped to induce mild symptoms and immunity. Vaccines still provide one of the best means for preventing infectious disease. Active research is ongoing to find better ways to suppress transplant rejection and to boost immunity to tumors.

ESSENTIAL BACKGROUND

- **Immunogenicity (Chapters 1 and 2)**
- **Designer antibodies (Chapter 3)**
- **MHC protein inheritance and function (Chapter 5)**
- **Innate, humoral, and cellular immune effector mechanisms (Chapters 9 and 10)**
- **Hypersensitivities (Chapter 11)**
- **Regulation of the immune system and immunosuppressive agents (Chapter 12)**

TOPIC 1: VACCINES

KEY POINTS

✓ *What are the properties of a successful vaccine?*

✓ *What are the advantages and disadvantages of each major vaccines type?*

✓ *Under what circumstances is passive immunization necessary or desirable?*

Vaccination involves deliberate exposure to antigen under conditions where disease should not result. The vaccine should be of the proper dose and chemical characteristics to stimulate protective immunity, which may involve secretion of neutralizing antibody or production of memory cytotoxic T lymphocytes (CTL) or T_H1 cells. A good vaccine should not induce autoimmunity or hypersensitivity. Ideally, it will be inexpensive to produce, store and administer.

The Salk polio vaccine is an example of an **inactivated (killed) virus vaccine** made by growing virulent polio virus in tissue culture and then treating it with formaldehyde so that it cannot replicate. Polio virus does not undergo rapid mutation, so a mixture of three serotypes is sufficient to generate neutralizing antibody that provides protection against all common polio viruses. Advantages of this vaccine are a low risk of infection, safety for people whose immune systems are compromised, and stimulation of immunity to antigens in their natural conformation. Disadvantages are that because the virus cannot replicate, a large dose is required to stimulate immunity; periodic boosters must be given to maintain immunity; only humoral immunity is induced; and administration by injection is costly. Influenza virus vaccine is another inactivated virus vaccine. It has the additional disadvantage that flu viruses mutate rapidly, so that new antigenic specificities must be included in each year's vaccine. Predictions are made in the spring about which antigens will predominate the following year; if those predictions are wrong, the vaccine is not protective.

The Sabin oral polio vaccine and the measles, mumps, and rubella vaccine are examples of **attenuated virus vaccines**, made less virulent by growth in animals or tissue culture. Advantages of attenuated vaccines are generation of memory cellular and humoral responses, lower dose requirements, and native antigen conformation. The Sabin vaccine is administered orally, which is less expensive and induces mucosal immunity and IgA synthesis. Disadvantages of attenuated vaccines are possible reversion to virulence and the potential to cause disease in immunosuppressed people.

Subunit vaccines contain purified antigens or peptides rather than whole organisms; an example is *Bordatella pertussis* antigens included in the acellular diphtheria-pertussis-tetanus (DPT) vaccine. Subunit vaccines are not infectious and they are less likely to induce unfavorable immune reactions that may cause side effects, such as the fever and irritability that were common and the neurological complications that were rare side effects of the whole pertussis vaccine. Subunit vaccines may be less immunogenic than whole organism vaccines. Other protein vaccines that induce good protective immunity are the diphtheria and tetanus **toxoid** components of DPT: toxins treated to eliminate their toxicity but still able to induce neutralizing antibodies. Other successful subunit vaccines are *Haemophilus influenzal B* (HiB) and *Streptococcus pneumoniae* capsular polysaccharides. HiB polysaccharide is complexed with diphtheria toxoid to increase its immunogenicity in infants. A new approach to developing vaccines to parasites is to isolate parasite peptides from host cell major histocompatibility complex (MHC) and use those peptides (synthesized in bulk in the laboratory) to induce immunity. Peptide immunogenicity can be increased by giving them in **ISCOMs** (**immunostimulating complexes**), lipid micelles that can transport the peptides into the cytoplasm of dendritic cells for presentation on MHC class I.

Recombinant vaccines are those in which genes for the desired antigens are expressed in bacteria and injected as subunit vaccines or inserted into a nonvirulent vector for a whole-organism vaccine. The vector should be chosen to be safe and also easy to grow and store, reducing production costs; antigens that do not elicit protective immunity or that elicit damaging responses can be eliminated from the vaccine. A disadvantage of recombinant vaccines is their cost to develop. The only recombinant vaccine currently in human use is hepatitis B virus (HBV) vaccine, a recombinant subunit vaccine. HBV surface antigen is produced from a gene transfected into yeast cells and purified for injection. This is much safer than using HBV, which causes hepatitis and liver cancer.

DNA vaccines are the newest vaccines and also the least well understood, although to date they are effective and safe. Genes for the desired antigens are located, cloned, and injected using compressed gas into muscle cells that express pathogen DNA to stimulate the immune system. Both humoral and cellular immunity are induced.

Passive immunization is used when there is not enough time to induce active protective immunity or prophylactically in certain immune deficiencies. Serum containing specific antibody from a human or animal in response to vaccination or environmental exposure to antigen provides passive humoral immunity. Examples include anti-venom for snake bite, Rhogam (human anti-Rh) to block formation of IgG anti-Rh antibodies in Rh^- mothers to Rh^+ red blood cells from their infants, and human gamma globulin given to children who have humoral immune deficiencies. Passive immunization confers adequate protection in the short term but has a half-life of about 60 days. When the antibodies come from another species, serum sickness may result.

Topic Test 1: Vaccines

True/False

1. An example of passive immunization is a vaccine composed of killed pathogen.
2. DNA vaccination induces cellular but not humoral immunity.

Multiple Choice

3. To make a recombinant vaccine to a virus, the virus DNA is expressed
   a. Covalently linked with a polysaccharide.
   b. In a genetic variant of the same virus.
   c. In a less virulent vector, which may be a virus or a bacterium.
   d. In muscle cells of the immunized person.
   e. On an ISCOM.
4. An advantage of passive immunization is that it provides __________ immunity.
   a. Cellular
   b. Faster
   c. Longer lasting
   d. More specific
   e. Safer

Short Answer

5. Why must the dose of antigen be higher in a killed vaccine than in an attenuated vaccine?
6. What measures can be taken to increase the immunogenicity of a vaccine?

Topic Test 1: Answers

1. **False.** An example is transfer of human or animal serum containing antibody to pathogens or toxins: anti-snake venom, anti-tetanus toxin, human gamma globulin (containing antibodies to a variety of common human pathogens).
2. **False.** DNA vaccination induces both cellular and humoral immunity.
3. **c.** Virus DNA is expressed in a virus, a bacterium, or a yeast.
4. **b.** Passive immunization provides faster but short-lived humoral immunity. It is safer for people who might not be able to make an active immune response.

5. Pathogens comprising an attenuated vaccine multiply to provide additional antigenic stimulation, whereas those in the killed vaccine cannot.
6. Vaccine immunogenicity is increased by giving the antigen with adjuvants or ISCOMs, higher antigen doses, and additional boosters. Covalently linking protein carriers to polysaccharides increases their immunogenicity, especially in children. Attenuated pathogens and DNA vaccines induce both cellular and humoral immunity, and oral vaccine delivery targets immune effectors to the site of pathogen entry.

TOPIC 2: TRANSPLANTATION

KEY POINTS

✓ *Which MHC antigens are most important to match for transplant compatibility?*

✓ *How are donors and recipients matched for MHC class I and MHC class II?*

✓ *By what immune mechanisms are organ transplants rejected?*

✓ *What special problems are encountered in bone marrow transplantation?*

✓ *What immunosuppressive agents are used to prevent transplant rejection?*

Recognition of foreign MHC antigens by T cells and their activation to effector T_C or T_H leads to transplant rejection. Autografts (skin transplanted from one location to another) and grafts from an identical twin have no foreign MHC and require no antirejection drugs. Allografts come from members of the same species who may have different MHC alleles. Siblings have a one in four chance of sharing all MHC antigens. Close relatives usually share more MHC antigens than unrelated people, and people from different parts of the world share fewer MHC antigens. Xenografts come from different species; their MHC may be too foreign to be recognized by T cells, but adhesion molecules and surface carbohydrates to which humans have antibodies induce rapid rejection.

Tissue typing involves identifying MHC antigens on donor and recipient cells and then matching as many alleles as possible. Differences in class I [especially human leukocyte antigen (HLA)-B] and class II HLA-DR alleles contribute more to rejection than differences at other MHC loci; minor histocompatibility antigens are less important. Blood type antigens, present on many tissues, must be matched; in the case of transplanted hearts, which cannot be kept healthy for long outside the body, blood type antigens are usually the only ones matched. MHC antigens are typed using serology (anti-HLA antibodies), cell-mediated lysis (^{51}Cr release from donor targets by recipient T_C), or mixed lymphocyte reactions (^{3}H-thymidine uptake by recipient T_H or T_C in response to donor MHC). Serological assays are the fastest but cannot detect MHC antigens to which no antibody is available or measure how "foreign" different alleles may appear to the recipient's T cells.

Hyperacute graft rejection occurs immediately upon transplantation due to preformed antibodies, either natural antibodies to blood type antigens or anti-MHC antibodies formed in response to blood transfusions or during pregnancy. Hyperacute rejection fatally damages the organ and cannot be reversed; the only treatment is immediate graft removal. Acute graft rejection occurs days or weeks after transplantation. Symptoms include fever, skin rash, impaired organ function, and a mononuclear (T cell) infiltrate visible on biopsy. **Acute rejection** is usually due to cellular immunity to foreign MHC and is treated with antibody to intercellular adhesion molecule–1 (ICAM–1) and CD3 and the immunosuppressant drugs prednisone, azathioprine, and cyclosporin A. Immunosuppressant doses must be high enough to prevent rejection but low enough to permit the immune system to fight infection. Acute rejection also occurs due to foreign peptides presented on self MHC alleles, so it can occur in grafts between identical twins. After many years the recipient's immune system may come to view the graft as self and antirejection drugs can be reduced or discontinued.

In bone marrow transplantation, the hematopoietic system of the recipient is completely destroyed by irradiation or chemicals to make room for the transplanted marrow and to kill cancer cells. Rejection causes **graft-versus-host disease** (**GVHD**), where immune effector cells in donor marrow reject host tissues. Donor marrow treatment with anti-CD3, anti-CD4, and CD-8 kills mature T cells and reduces the incidence of GVHD. Placental cord blood contains many hematopoietic stem cells and few mature T cells and can replace bone marrow as a source of hematopoietic cell transplants.

The fetus is an almost perfect allograft. Most pregnancies are not rejected even though half the infant's antigens are foreign to the mother (or all in the case of surrogate pregnancy) and fetal cells contact maternal cells in the placenta and enter the mother's circulation during pregnancy. Suppression of the mother's immune response to the developing fetus's paternal antigen may be regulated by a combination of suppressor cells, blocking antibodies, nonclassical MHC in the placenta, nonspecific immunosuppressive molecules, and hormonal mediators.

Topic Test 2: Transplantation

True/False

1. Transplanted corneas do not have to be MHC matched because they have no circulation and recipient T cells cannot reach the graft.
2. Xenografts are transplants between genetically different members of the same species.

Multiple Choice

3. John Doe needs a kidney transplant. The best match for John would probably be
   a. An unrelated cadaver.
   b. His brother.
   c. His cousin.
   d. His mother.
   e. His wife.
4. One of the ways allograft rejection is prevented is by administration of
   a. Antibodies to CD3.
   b. Antibody to the foreign MHC.
   c. Interleukin (IL)-2.
   d. IgE.
   e. Rhogam.

Short Answer

5. How is rejection different for bone marrow transplants than for organ transplants?
6. Why is it important to match blood types between transplant donors and recipients?

Topic Test 2: Answers

1. **True**. Transplanted corneas are immunoprivileged tissues without contact with the immune system.
2. **False**. Xenografts are tissue transplants done between members of different species.
3. **b.** Siblings have a one in four chance of sharing all their MHC genes (and a two in four chance of sharing half). Parent–child matches will be at least 50% identical. Cousins may or may not be similar, whereas unrelated people have the lowest chance of matching.
4. **a.** Antibodies to the T cell CD3 block their activation to foreign MHC.
5. GVHD occurs in bone marrow rejection because the host has no immune system with which to reject the graft.
6. Preformed antibodies to blood group antigens exist in people with blood types A (anti-B), B (anti-A), and O (anti-A and anti-B) and cause hyperacute graft rejection of a graft with one of those antigens (which are on other tissues besides red blood cells).

TOPIC 3: TUMOR IMMUNOLOGY

KEY POINTS

✓ *How does the immune system detect tumors?*

✓ *What specific and nonspecific immune responses might eliminate tumors?*

✓ *How do tumors evade the immune system?*

✓ *How is immunology used to diagnose and treat cancer?*

✓ *How can gene therapy be used to stimulate the immune response to cancer?*

Cancer cells undergo unregulated cell proliferation and metastasize to distant sites. Cancer results from several factors, including genetic predisposition, viral transformation, environmental mutagens, and tumor promoters. The theory of immune surveillance is that the immune system continually recognizes and eliminates tumor cells; when a tumor escapes immune surveillance and grows too large for the immune system to kill, cancer is the result.

Immunogenic tumors have tumor-specific antigens on their surface, antigens not present on nontumor cells; many tumor-specific antigens are viral proteins from a tumor-causing virus. T_C recognize tumor-specific antigens presented on self class I and kill the tumor; natural killer cells also kill tumor cells. Tumor-associated antigens shared with normal cells are more common and less immunogenic. Immunity to them may be suppressed because they are considered "self"; if immunity is induced, it may damage normal tissue. Some tumors reduce their MHC expression; many lack co-stimulatory molecules or adhesion molecules required for them to bind T_C. Tumors may actively suppress cellular immunity by producing transforming growth factor beta; some tumors, including myeloma and HTLV-1 T-cell leukemia, produce cytokines that stimulate their own growth. Antibodies to tumor antigens may promote tumor survival (enhancing antibodies) if they bind without being cytotoxic, hiding tumor antigens from T cells and inducing the tumor to reduce antigen expression. Tumors also change or shed their antigens spontaneously.

Radioisotopically labeled mouse monoclonal antibodies to tumor-specific and tumor-associated antigens are used to detect tumors and monitoring tumor therapy. Tumor-specific antibodies linked to toxins, antitumor drugs, or very energetic radioisotopes are used to kill tumor cells. Problems with immunotherapy include the time and expense required to make a unique antibody for each tumor, lack of antibody access to the tumor interior, and HAMA responses to xenogeneic monoclonal antibodies. Production of human monoclonals has been difficult, but chimeric antibodies and humanized antibodies can be made that are less immunogenic. Tumor immunotherapy is not done until conventional therapies like surgery, chemotherapy, and radiation have been tried; surgically debulking the tumor is essential for success of immunotherapy, but radiation and chemotherapy induce immune deficiency.

Cytokines are used clinically to treat tumors. IL-2, interferon alpha, and tumor necrosis factor alpha are used in vivo, and IL-2 is used in vitro to expand populations of lymphokine-activated killer cells and tumor-infiltrating lymphocytes so they can be returned to the patient to kill tumor cells. Limitations on therapeutic use of cytokines include the need for locally high concentrations, their short biological half-lives, and unwanted side effects such as vascular shock and psychoses. Introduction of inactivated tumor cells transfected with genes for cytokines that stimulate immunity or antisense genes for inhibitory cytokines are undergoing clinical trials. For example, a granulocyte-macrophage colony-stimulating factor gene transfected into tumor cells recruits hematopoietic precursors and induces them to become dendritic cells, which can take up and present tumor antigens to T cells.

Topic Test 3: Tumor Immunology

True/False

1. Enhancing antibodies enhance the ability of CTL to kill tumor cells.

2. Tumor cells do not produce cytokines but can be induced to do so by transfecting them with cytokine genes.

Multiple Choice

3. Tumor-associated antigens are *not*
   a. Expressed on nontumor cells.
   b. Overexpressed self antigens.
   c. Presented on MHC class I.
   d. Shared with fetal cells (oncofetal antigens).
   e. Tumor specific.

4. Tumors do *not* escape immune surveillance by
   a. Downregulating their expression of MHC class I.
   b. Expressing antigens not found on other self tissues.
   c. Failing to express B7.
   d. Producing immunosuppressive cytokines like transforming growth factor beta.
   e. Shedding their membrane antigens.

Short Answer

5. Why are antitumor drugs immunosuppressive?
6. Why are there such serious side effects from cytokines given systemically?

Topic Test 3: Answers

1. **False.** Enhancing antibodies enhance tumor cell escape from T-cell recognition by covering tumor antigens and promoting their reduced expression and shedding.
2. **False.** Some tumors produce cytokines that stimulate their own growth or suppress immunity; they can be transfected with genes for immunostimulatory cytokines.
3. **e.** Tumor-associated antigens are expressed on normal cells and may be overexpressed self or oncofetal antigens. Like other self, they are presented in class I.
4. **b.** Tumors escape immune surveillance by all of the mechanisms except for expressing tumor-specific antigens, which makes them more immunogenic.
5. Antitumor drugs are immunosuppressive because they kill rapidly dividing cells, which include lymphocytes and other hematopoietic cells.
6. Cytokines usually function at very low local concentrations (in the picogram range). When they are given systemically at the high doses needed to be effective, they cause life-threatening shock similar to that seen in toxic shock syndrome.

CLINICAL CORRELATION: PIG PARTS FOR PEOPLE (XENOTRANSPLANTATION)

Because of the shortage of human organs for transplantation, there has been an increasing interest in transplanting animal organs. Pigs are inexpensive, easy to breed, and already raised for food; their organs are similar in size to human organs. Pig heart valves and skin are already transplanted to people. Skin serves as a temporary covering for severely burned patients but is rejected; heart valves are not rejected because living pig cells are removed before transplantation of cartilage. Xenograft rejection occurs by hyperacute mechanisms; preformed human antibodies to pig carbohydrates and adhesion molecules activate complement-mediated inflammation and tissue destruction. Pig complement-inhibiting molecules like decay accelerating factor (DAF) do not block human complement function. Transgenic pigs have been created that have five human genes: CD46, CD55, CD59, DAF, and H-transferase. The first four of these encode human complement-inhibitory proteins; H-transferase changes a pig surface carbohydrate residue to a human one. Transgenic pig organs survive for 30 hours in baboons instead of 1 hour, but xenografting from pigs to humans is not yet practical. Possible transfer of potentially lethal viruses from animals to humans via xenografts must also be prevented.

DEMONSTRATION PROBLEM

Melanomas are fast-growing skin tumors that also metastasize to distant locations in the body. They express tumor-associated antigens MAGE-1 and MAGE-3, which are also found on testicular tissue. How could you use monoclonal antibodies to MAGE to kill metastatic cancer? Include a description of the side effects of this therapy and why it might not be successful.

SOLUTION

Monoclonal antibodies to MAGE could be conjugated with antitumor drugs, radioisotopes, or toxins to deliver these cytotoxic molecules to melanoma metastases. Side effects could include damage to testicular tissue, which also expresses MAGE. If the monoclonal antibodies are not human, a HAMA response will reduce their effectiveness with each use; if they are humanized, they may retain their effectiveness longer.

Chapter Test

True/False

1. Annual influenza vaccines are required because the flu shot is a subunit vaccine.
2. Humanized monoclonal antibodies to tumor antigens are preferred over mouse monoclonals because they are more specific for human tumor antigens.
3. Transplants between identical twins do not require antirejection therapy.
4. Like pathogens, tumors can evade the immune system by changing their antigens.
5. Xenogeneic transplants are usually rejected quickly because their MHC alleles are so foreign to human T cells.

Multiple Choice

6. A virus molecule expressed on the membrane of a virus-induced tumor cell is an example of a(n)
   a. Angiogenic molecule.
   b. Carcinoembryonic antigen.
   c. Proto-oncogene.
   d. Tumor-associated antigen.
   e. Tumor-specific antigen.
7. Hepatitis B surface antigen vaccine, a recombinant vaccine produced in yeast cells,
   a. Cannot be given to immunosuppressed children.
   b. Uses the yeast cells as an adjuvant to increase its immunogenicity.
   c. Is less safe to administer than killed HBV vaccine.
   d. Is more immunogenic than the whole hepatitis virus would be.
   e. Requires several boosters because it is also a subunit vaccine.
8. Tumor cells do *not* evade the immune system by
   a. Downregulating expression of self MHC class I.
   b. Failing to express B7.

c. Inducing production of enhancing antibodies.
d. Secreting interferon gamma.
e. Shedding their surface antigens.

9. Hyperacute graft rejection
   a. Always occurs if HLA-B or HLA-DR alleles are mismatched.
   b. Can be controlled with cyclosporin A and anti-CD3.
   c. Causes most transplantation failures.
   d. Is due to the presence of antibodies against graft surface antigens.
   e. Is due to T-cell recognition of more than three differences in MHC alleles.

10. Passive immunization to a bacterial toxin would be recommended if
   a. Exposure to the toxin had already occurred in an unimmunized person.
   b. The person had been recently immunized with the toxoid.
   c. The toxin could not reproduce in the body.
   d. The toxin was not immunogenic.
   e. The toxoid was available only as part of a conjugate vaccine.

Short Answer

11. Why is acellular (subunit) pertussis vaccine an improvement over the whole organism vaccine?

12. What are minor histocompatibility antigens and how can they cause graft rejection?

13. How can bone marrow cells be treated to decrease the chances of GVHD?

14. What problems are associated with giving cytokines systemically to stimulate antitumor immunity?

Essay

15. What problems does human immunodeficiency virus (HIV) pose for vaccine development?

Chapter Test

1. **F** 2. **F** 3. **T** 4. **T** 5. **F** 6. **e** 7. **e** 8. **d** 9. **d** 10. **a**

11. Killed whole *B. pertussis* vaccine often induced fever and restlessness and more rarely caused serious neurological side effects not caused by the acellular vaccine.

12. Minor histocompatibility antigens are tissue or sex specific or are foreign peptides presented on self MHC. They are visible to T cells and can activate them to initiate graft rejection, even in grafts where all MHC are matched.

13. GVHD is due to mature T cells in the graft that recognize the recipient as foreign. The graft can be treated with antibodies to CD3, CD4, and CD8 plus complement to kill mature T cells and reduce the risk of GVHD.

14. High systemic doses of cytokines like IL-2 induce life-threatening vascular shock and psychotic reactions.

15. HIV causes lethal infections, so safety is a prime concern. People infected with HIV make vigorous humoral and cellular immune responses, but the virus still eventually kills so many $CD4^+$ T cells that they die of infection. It is not clear whether cellular or humoral immunity is more protective against HIV, and the virus is often transmitted from person to person inside cells, shielding it from neutralizing antibodies. HIV rapidly mutates its surface antigens, so immunity that is developed to the infecting serotype becomes ineffective. Immune effector mechanisms that kill infected cells increase loss of T_H, and HIV suppresses the host's ability to replace lost T cells even when $CD4^+$ counts are still high.

Check Your Performance:

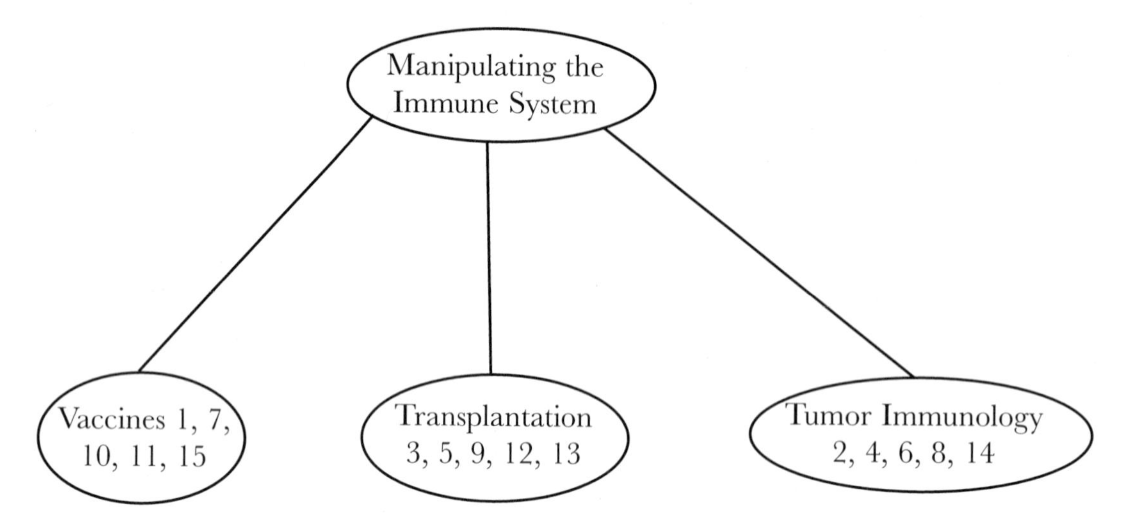

Check your understanding of this chapter by noting the number of questions for each topic you missed on the chapter test.

Final Exam

True/False

1. Chemokines activate complement to attract leukocytes to the infection site.
2. Viruses cannot be lysed by complement.
3. Both location and activation state influence leukocytes extravasation.
4. Phagocytes can release their proteases and oxygen radicals to kill microbes they have not phagocytosed.
5. Each epitope-specific B cell can activate only T cells specific for the same epitope.
6. If a mature cytotoxic T lymphocyte (T_C) encounters a peptide it does not recognize on class I, it will die.
7. An example of hyperacute graft rejection would be natural anti-B antibodies in the recipient binding B antigen on the donor organ.
8. Autoimmune diseases are normally treated by injecting a monoclonal antibody to the self antigen to prevent its sensitization of T and B cells.
9. Graft-versus-host responses occur when mature donor T cells become activated by recipient major histocompatibility complex (MHC) or minor histocompatibility antigens.
10. Passive immunization against bacterial pathogens would be favored over active immunization in a child who lacked the ability to synthesize antibody.

Multiple Choice

11. Cytokines do *not*
    a. Activate cells specifically by binding antigen co-receptors.
    b. Activate cells far away from the cell that produced them.
    c. Bind soluble cytokine receptor molecules.
    d. Bind to the cell that synthesized them.
    e. Get synthesized de novo in response to cell-activation signals.

12. Complement
    a. Activation occurs after a lag phase.
    b. Activation requires acute phase protein synthesis.
    c. Increases in concentration during an immune response.
    d. Is both activated and inactivated by proteolytic enzymes.
    e. Is synthesized in response to innate pathogens.

13. Inflammatory cytokines produced by macrophages do *not* activate
    a. B cells to secrete acute phase proteins.
    b. Integrin on leukocytes to bind more strongly to vascular cell adhesion molecules (CAMs).
    c. Neutrophils to be more cytotoxic.
    d. Natural killer (NK) cells to kill virus-infected cells.
    e. Vascular endothelium to upregulate CAMs.

14. If a mature T_C cell binds peptide presented on class I of a target cell without membrane B7, the T_C cell will
    a. Be activated, but not as strongly as if B7 were present.
    b. Be activated if cytokines from a nearby T_H1 cell can provide co-stimulation.
    c. Induce B7 expression on the target cell.
    d. Proliferate but not differentiate.
    e. Undergo apoptosis.

15. Humoral immunity does *not* involve
    a. Antibody-dependent cell-mediated cytotoxicity by NK cells.
    b. Antibody neutralization of viruses.
    c. Antibody secretion by plasma cells.
    d. Classical complement activation.
    e. Delayed-type hypersensitivity (DTH) by macrophages.

16. B cells are
    a. Negatively selected in primary lymphoid organs after foreign antigen binding.
    b. Negatively selected in secondary lymphoid organs after self-antigen binding.
    c. Positively selected in primary lymphoid organs after somatic recombination.
    d. Positively selected in secondary lymphoid tissues after somatic hypermutation.
    e. Rendered anergic in the circulation by contact with haptens.

17. Granulomas form in the lungs of patients with tuberculosis. This is an example of
    a. An autoimmune disease.
    b. A primary immune response.
    c. DTH.
    d. Transformed cells forming a metastatic tumor.
    e. Type I hypersensitivity.

18. A state of tolerance is favored over an immune response if antigen is administered
    a. Before puberty.
    b. In the presence of mycobacteria, which divert the T-cell and macrophage response.
    c. To IgM^+ IgD^- B cells.
    d. To T cells before T-cell receptor rearrangement.
    e. When the recipient is anesthetized.

19. Tumor-associated antigens
    a. Are expressed only on cancer cells.
    b. Are found predominantly on lymphoid tumors.
    c. Are virus proteins.
    d. Can be measured as an indicator of cancer therapy success.
    e. Help the cancer cell suppress the immune system.

20. Memory T cells would *not* be produced after
    a. Administration of oral polio vaccine.
    b. Infection with *Haemophilus influenzae* (a bacterium).
    c. Infection with influenza virus.
    d. Injection of a DNA vaccine to human immunodeficiency virus.
    e. Injection of pneumococcal polysaccharide.

Short Answer

21. What is happening during the first few days of a primary immune response to a virus infection in the mucous membranes of the nose before a rise in serum antibody can be detected?

22. Pat went to study at a friend's apartment without knowing that the friend had a cat, to which Pat was severely allergic. Within minutes, Pat was wheezing; he took his antihistamines and made a date to study at the library later. He felt fine the rest of the evening, but about two in the morning he awoke with difficulty breathing. What caused Pat's symptoms?

23. Describe two ways in which a tumor can avoid immune elimination.

24. Name two immune deficiencies that could result in an increased susceptibility to bacterial infections and explain why susceptibility would increase.

Essay

25. Cindy, a 5-year-old child who has just finished a round of chemotherapy for leukemia, is due for her booster DPT, oral polio vaccine, measles, mumps, and rubella (MMR), and hepatitis B vaccinations. She has three school-aged siblings. What would be your recommendation to her parents about her vaccinations? Explain the reasons for your recommendations.

Final Exam: Answers

1. **F** 2. **F** 3. **T** 4. **T** 5. **F** 6. **F** 7. **T** 8. **F** 9. **T** 10. **T** 11. **a** 12. **d**

13. **a** 14. **b** 15. **e** 16. **c** 17. **c** 18. **c** 19. **d** 20. **e**

21. The virus infects and kills nasal cells. Extracellular virus is phagocytosed and destroyed and may activate complement to promote inflammation. Interferon (IFN) alpha and IFN beta are secreted and bind to host cell membranes to activate inhibitors of virus replication They also increase expression of class I, TAP, and proteasome components to increase antigen presentation to T_C. NK cells are activated by IFN alpha, IFN beta, and IL-12 to kill virus-infected cells and by IL-12 and tumor necrosis factor alpha to produce IFN gamma, a strong macrophage activator. As virus protein is presented to T_H and T_C cells and B cells are activated, the adaptive immune response begins.

22. Pat was probably suffering from a late-phase type I hypersensitivity response to the cat dander, due to mast cell synthesis of chemokines and platelet-activating factors to attract leukocytes, cytokines (including IL-4) to activate eosinophils and stimulate their synthesis in the marrow, and leukotrienes (SRS-A) to promote blood flow, smooth muscle constriction, and mucus secretion. The same antigen-IgE-FcεR stimulus that caused Pat's immediate hypersensitivity also stimulated the late-phase response.

23. Tumor cells escape the immune system by not expressing tumor-specific antigens, by suppressing MHC expression (and therefore tumor-peptide presentation), by shedding or altering their antigens, or by secreting immunosuppressive cytokines.

24. Deficiencies most likely to result in increased bacterial infections are those that affect complement components, C-reactive protein (CRP) and mannose-binding protein (MBP),

macrophages and neutrophils (including defects in phagosomes acidification, nitric oxide, and oxidative burst, and exogenous antigen presentation on class II), B cells and antibody production, and helper T cells.

25. Cancer chemotherapy suppresses the immune system, so Cindy is at increased risk of infection but may not respond optimally to vaccines. DPT vaccine is composed of tetanus and diphtheria toxoids and a subunit of *B. pertussis*, and hepatitis B is a recombinant subunit vaccine, so they would be safe to give. Oral polio vaccine and MMR are attenuated viruses, which could cause serious disease in an immunosuppressed child; killed polio vaccine and killed MMR should be substituted, even if Cindy needs additional boosters later to maintain her immunity. If Cindy were seriously immunosuppressed, human gamma globulin could be used to provide passive immunization.

INDEX

active immunity, 12
acute phase protein, 116
adaptive immunity, 4, 17, 24, 116, 121–137, 150, 155
ADCC, 10, 127, 142
adhesion molecule, 111–113, 123–124, 126, 140, 154, 167
adjuvant, 6, 12, 16
adoptive transfer, 24
affinity, 36
affinity chromatography, 19, 21
affinity constant, 36
affinity maturation, 126
agglutination, 18, 142
AIDS, 157–158
alkylosing spondylitis, 65, 153
allergen, 139
allergy, 10, 16, 38, 139–149
allogeneic, 64, 77
allograft, 145, 165
alloreactive T cells, 77
allotype, 38
alternative complement pathway, 100, 103, 116
alternative mRNA splicing, 52
anamnestic response, 126
anaphylatoxin, 101, 103
anchor residue, 58
anergy, 50, 123
angioneurotic edema, 107
antibody, 4, 10, 12, 17–19, 22, 33–44, 123, 150, 153, 163, 166
antigen, 5, 17, 22, 150, 153, 162, 167
antigen binding, 6, 22, 35, 50, 58, 123, 163
antigen processing, 62–63, 125
antigenic drift, 156
antigenic shift, 156
antigenic variation, 156
Ap-1, 124
APC, 6, 24, 60, 62, 64, 77, 123, 125, 127, 129
apoptosis, 8, 75, 126, 128–129, 156
appendix, 8
Arthus reaction, 143
ascites tumor, 24
atopy, 139
attenuated vaccine, 4, 163
autograft, 165
autoimmunity, 152–154, 162
autoradiography, 21
avidity, 36, 70

B cell, 4, 6, 21–24, 50, 57, 94, 101, 121, 123, 150–151
B-cell activation, 121–122, 125, 140
B-cell development, 50
b_2M, 57
B7, 123, 129
bare lymphocyte syndrome, 64
basophil, 3
BCR, 6, 10, 38, 61, 70, 122, 125
Bence Jones protein, 41
Blk, 125
bone marrow, 8, 50
Btk, 53, 155

C1 INH, 105, 107
C3 convertase, 103
C4BP, 105
C5 convertase, 103
calnexin, 63
calreticulin, 63
CAM, 90, 111, 116
carbohydrate antigen, 16, 122
carrier protein, 6, 16, 125
CD antigen, 8
CD2, 8, 74, 124
CD3, 8, 71, 74, 124, 153, 166
CD4, 8, 58, 64, 70, 74, 157, 166
CD5, 68
CD8, 8, 40, 58, 64, 70, 74, 124, 128, 166
CD11b/18, 101
CD14, 114
CD19, 8
CD21, 8
CD25, 75
CD28, 123
CD34, 50, 112
CD40, 93, 126, 129
CD40L, 126, 129
CD46, 105, 169
CD55, 105, 169
CD59, 105, 169
CD115, 8
CDR, 33, 45, 70
cell adhesion molecule (CAM), 111
cell-mediated lysis, 22
cellular immunity, 4, 17, 124, 128–130, 163
central lymphoid organ, 8
CGD, 118
chaperone, 63
chemokine, 89, 92, 113–114
chemotaxis, 114
chimeric antibody, 40
^{51}Cr release, 22, 165
chronic desensitization, 140
chronic granulomatous disease, 118
class I MHC, 6, 10, 58–59, 71, 116, 123, 155, 163, 167
class II MHC, 6, 8, 11, 58–59, 60, 71, 116, 121, 123, 125, 155
classical complement pathway, 100, 103
CLIP, 62
clonal deletion, 50, 151
clonal selection, 4
co-stimulation, 121–123, 150–151
combinatorial diversity, 48, 73
complement, 4, 10, 17, 24, 60, 100–110, 113, 142, 150, 153

complement fixation, 19, 66
complement regulation, 105–106
complement split fragment, 103
congenic mice, 24
Coombs test, 20
CR, 6, 8, 10, 101, 105, 143
cross-reactivity, 6, 36, 77
CRP, 116
CTLA, 4 123
cyclosporin A, 154
cytokine, 4, 10, 17, 22, 60–61, 78, 89–99, 112, 116, 124, 129, 140, 150, 156, 167
cytokine antagonist, 89, 94
cytokine pleiotropism, 89
cytokine receptor, 93–94
cytokine redundancy, 89
cytokine synergism, 89
cytosolic processing pathway, 63
cytotoxic T cell, 4, 8, 10, 17, 22, 40, 61, 63–64, 66, 75, 90, 124, 128, 153

DAF, 105–106, 169
dendritic cell, 6, 57, 60, 123, 168
DNA vaccines, 163
double-negative T cell, 75
double-positive T cell, 75
DTH, 90, 129, 145

early induced response, 116
EBV, 94, 101
eczema, 140
effector cell, 3, 123–124
ELISA, 19, 22, 106, 157
endogenous antigen, 10, 17, 57, 123, 156
endoplasmic reticulum, 59
endosomal processing pathway, 62
eosinophil, 3
epitope, 6, 36, 122, 125–126
equilibrium dialysis, 36
erythroblastosis fetalis, 147
exogenous antigen, 10, 17, 57, 62, 123, 156
experimental autoimmune encephalitis, 153
extravasation, 101, 112, 114

Fab, 34
FACS, 21
Factor D, 106
Factor H, 105–106
Factor I, 105–106
Fas, 94, 129, 156
FasL, 129
Fc, 34
FcR, 6, 10, 60, 140–143
FDC, 101, 126–127, 158
fever, 17, 78, 116, 163
Ficoll Hypaque, 21
FK506, 154
flow cytometry, 21
fluorescence-activated cell sorting, 21
fluorochrome, 21
follicle, 50, 126
framework region, 33, 45
Fv antibody, 40
Fyn, 124–125

gamma delta T cells, 75
gamma globulin, 33
germinal center, 126
GlyCAM, 112
GM-CSF, 90, 92, 94
Goodpasture's syndrome, 153
graft rejection, 65, 77, 142, 145, 165
granulocyte, 3, 21, 23–24, 155
granzyme, 127–128
Grave's disease, 153
GVHD, 166

H-2, 57
HAMA response, 40, 168
haplotype, 59
hapten, 6, 125
helper T cell, 4, 10
hematopoiesis, 8, 17
hemolysis, 20
heteroconjugate, 40
HEV, 112
hinge, 34
histamine, 38, 139
HIV, 156–157
HLA, 57–59, 152–153, 165

HLA-DM, 62
HRF, 105–106
humanized antibody, 40
humoral immunity, 4, 124–127, 163
hybridoma, 24
hyper IgM syndrome, 131
hyperacute graft rejection, 142, 165
hypersensitivity, 139–149
hypervariable region, 33

ICAM, 112, 166
IDDM, 65, 153
idiopathic hemochromatosis, 65
idiotype, 38, 151
IFN alpha, 90, 92, 116, 168
IFN beta, 90, 92, 116, 145
IFN gamma, 59, 63, 90, 92, 96, 116, 126, 129, 145
Ig alpha/Ig beta, 50, 125
Ig constant region, 33, 45
Ig domain, 33, 71
Ig genes, 45–56
Ig isotype, 10
Ig superfamily, 57
Ig variable region, 33, 45
IgA, 10, 16, 34, 36, 38, 52, 121, 126, 155–156, 163
IgD, 10, 16, 34, 36, 38, 52
IgE, 10, 16, 34, 36, 38, 52, 121, 126, 139–140
IgG, 10, 16–18, 34, 36, 38, 52, 100, 121, 126, 140–141, 143, 153, 164
IgM, 10, 16–18, 23, 34, 36, 38, 52, 100, 121–122, 126, 141, 143
IL-1, 90, 92, 116, 123, 131, 151
IL-2, 24, 90, 92, 94, 96, 123–124, 126, 131, 154, 157, 168
IL-2R, 75, 94, 124, 157
IL-3, 90, 92, 94, 129
IL-4, 90, 92, 94, 96, 126, 140
IL-5, 90, 92, 94, 96, 126
IL-6, 90, 92, 96, 116, 126
IL-7, 90, 92, 94
IL-8, 90, 92, 116
IL-9, 94
IL-10, 90, 92

IL-12, 90, 92, 118
IL-15, 94
immune complex, 101, 127, 151
immune deficiency, 23, 155–156
immune surveillance, 167
immunoblot, 19
immuno-electron microscopy, 21
immunoelectrophoresis, 19
immunofluorescence, 21
immunogen, 5, 16, 77, 163, 177
immunoglobulin, 10, 90
immunohistochemistry, 21
immunophilin, 154
immunotoxin, 40
inactivated vaccine, 162–163
inbred mice, 24
inflammation, 3, 17
inflammatory T cell, 61
innate immunity, 3, 17, 23, 111–120, 155
integrin, 101, 112–113
interdigitating dendritic cells, 60, 125
interleukin, 89
ISCOMS, 163
isotype, 37–38
isotype switch, 51, 121, 126
ITAM, 50, 71, 124–125

J chain, 37–38, 126
junctional diversity, 48, 73

K cell, 127
killed vaccine, 162
killer inhibitory receptors, 8, 116
knock-out mice, 24

late-phase response, 140
Lck, 71, 124–125
leprosy, 96
leukocyte, 3, 10, 24, 112, 140
leukocyte adhesion deficiency, 118
LFA-1, 112, 123, 128
limiting dilution, 24
LMP, 59, 63
LPAM-1, 112
LPS, 114, 122, 129
lymphatic vessel, 8
lymphocyte, 4, 21, 24, 150
lymphoid progenitor, 74
lymphokine, 89
lymphokine-activated killer cell, 168
lymph node, 8, 16, 111

MIIC vesicle, 62
M cell, 8
mAb, 24, 38, 40, 168
MAC, 101, 103
MAC-1, 101, 112, 113, 114
macrophage, 3, 6, 10, 16–17, 21, 23, 57, 60, 90, 92, 94, 96, 114, 116, 123, 129, 156
MAdCAM-1, 112
MAGE, 170
MBL, 116
MCP, 92, 130
membrane attack complex, 101, 103
membrane marker, 8, 21
memory cell, 4, 16, 121, 124, 126
memory response, 4, 126, 150, 163
MHC, 6, 22, 24, 57–69, 70, 77, 90, 123, 126, 150, 163, 166–167
MHC restriction, 64, 77
minor histocompatibility antigen, 65
MIP, 105
MIRL, 105
mitogen, 6
mixed lymphocyte reaction, 22
molecular mimicry, 65
monoclonal antibody, 24, 38, 40, 168
monocyte, 3
monokine, 89
mucosal associated lymphoid tissues, 8
multiple myeloma, 41
multiple sclerosis, 153
myeloma protein, 34, 40
myeloma tumor, 23

negative selection, 21, 50, 75, 123
neutrophil, 3
NFAT, 124, 154
NFκB, 124, 157
NK cell, 3, 10, 22–24, 116, 129, 155, 167
NO, 90
nude mice, 23

opsonin, 4, 101
Ouchterlony, 19
oxidative burst, 114,116

P selectin, 112
p150.95, 101, 112, 114
PAGE, 19
paramagnetic beads, 21
passive immunization, 12, 163
pathogen, 3
perforin, 127–128
peripheral lymphoid organ, 8, 123
Peyer's patch, 8
phagocyte, 3, 10, 114, 155
phagocytosis, 60, 114, 116, 129, 156
phagosome, 114
phospholipase C, 124
plasma cell, 4, 23, 126
pluripotent stem cell, 8
PMN, 3, 10
polyclonal activation, 6, 22, 122
polyclonal antibody, 24, 38
positive selection, 75
pre-B cell, 50
pre-B receptor, 50
preTα, 75
precipitation, 19
primary lymphoid organ, 8
primary response, 17, 126, 150
pro-B cell, 50
productive rearrangement, 48
properdin, 105–106
proteasome, 116
protectin, 105
protein kinase C, 124

radiation chimera, 24
radioimmunoassay, 19
RAG, 48, 52, 73
RANTES, 92, 94
Ras, 124

RAST, 140
RBC, 18–19, 21
receptor editing, 50
recombinant vaccine, 163
recombinase, 45, 50
recombination signal sequence, 48, 73
reverse transcriptase, 157
Rh antigen, 20, 164
Rhogam, 164
rheumatoid factor, 143
RIA, 19

scavenger receptor, 114
SCID, 24,155
second messenger, 124–125
secondary lymphoid organ, 8, 123
secondary response, 126
secretory component, 38
secretory IgA, 38
selectin, 111–112, 114
self-antigen, 50, 75, 123, 152
sensitization, 139
septic shock, 116
serotype, 19, 36, 162
serum, 18
shock, 78
signal transduction, 50, 71
single-positive T cell, 75
somatic hypermutation, 49, 73, 126
somatic recombination, 45, 48, 50
spleen, 8, 153
S protein, 105
SRS-A, 140
stem cell, 8, 74
subunit vaccine, 163
superantigen, 78
suppressor T cell, 140, 151
surrogate alpha chain, 75
surrogate light chain, 50
switch region sequence, 52
syncytia, 157
syngeneic, 64
systemic lupus erythematosis, 153

T cell, 4, 21–22, 24, 111–112, 145
T-cell activation, 123–124
T-cell development, 74–75
T-cell education, 64, 74–75
T-dependent antigen, 121–122, 126
T-independent antigen, 6, 121–122, 126
TAC, 94
TAP, 59, 63, 66, 116, 129, 155
tapasin, 63, 153
target cell, 6, 129
T_C cell, 4, 8, 10, 17, 22, 40, 61, 63–64, 66, 75, 90, 123–124, 128, 151, 153, 155, 162
TCR, 6, 58, 70–81, 123, 126, 157
TCR genes, 73
TGF beta, 90, 92
T_H cell, 4, 8, 11, 16–17, 22, 60–61, 64, 66, 75, 90, 92, 96, 116, 121, 123–126, 140, 155
T_H1 cell, 61, 90, 92, 96, 118, 123–124, 145, 151, 162
T_H2 cell, 61, 90, 92, 96, 123–124, 130, 151
thymus, 8, 23, 60, 74–75, 122, 150
titer, 19
TNF alpha, 90, 92, 116, 129–130, 140, 168
TNF beta, 90, 93, 129–130
tolerance, 16, 151
tonsils, 8
toxoid, 12, 163
transgenic mice, 24
transplantation, 165–166
trypan blue, 21
TSST, 78
tumor immunology, 167–168
tumor infiltrating lymphocyte, 168

urticaria, 140

vaccine, 162–164
vascular addressin, 111
vascular endothelium, 116
VCAM, 112
viability stain, 21
vitronectin, 105
VLA, 112
VpreB, 50

Western blot, 19

X-SCID, 94
xenograft, 165
xenotransplantation, 164
XLA, 53

ZAP-70, 124